PRAISE FOR
INVENTING FREEDOM

"Daniel Hannan's unbridled passion for limited government speaks to a generation shaken and very much stirred, but as he suggests in this book, it is very much a continuing part of history. Freedom and self-reliance are our very DNA. Hannan gets that, connects that, and historically recounts that."

—NEIL CAVUTO, FOX BUSINESS

"Daniel Hannan clearly outlines the need for English-speaking peoples to remember their heritage of democratic freedoms and institutions. His clarion call for action should be heeded by those of us around the world who share and value this heritage."

—PRESTON MANNING, FOUNDER OF THE REFORM PARTY OF CANADA AND FORMER LEADER
OF THE OPPOSITION

"This is a distinguished, perceptive, and helpful contribution to the political debate."

—PAUL JOHNSON, AUTHOR OF *A HISTORY OF THE AMERICAN PEOPLE* AND
A HISTORY OF THE JEWS

"Daniel Hannan has found the key to the success of the English-speaking peoples: the unique political and legal institutions that make us what we are. His book teaches us to keep fast to that legacy and, in our turn, to pass it intact to the next generation."

—ANDREW ROBERTS, AUTHOR OF *A HISTORY OF THE ENGLISH-SPEAKING
PEOPLES SINCE 1900*

"Daniel Hannan marries a life in politics with first-class analytical writing like few others. This is yet another work of wit, elegance, and outstanding insight."

—MICHAEL DOBBS, AUTHOR OF THE HOUSE OF CARDS TRILOGY AND
WINSTON'S WAR

"English speakers have invented many things, among the most important of which are not artifacts but the twin concepts of personal freedom under lawful, self-chosen government. Daniel Hannan's book explains in layman's language that these twin concepts—still everywhere under assault—are no flukes but derive specifically from an essentially English-speaking culture."

—FREDERICK FORSYTH, AUTHOR OF *THE DAY OF THE JACKAL* AND
THE ODESSA FILE

ALSO BY DANIEL HANNAN

The New Road to Serfdom: A Letter of Warning to America

INVENTING
FREE

How the English-Speaking
Peoples Made the Modern World

DOM

DANIEL HANNAN

BROADSIDE BOOKS
An Imprint of HarperCollins*Publishers*

HarperCollins books may be purchased for educational, business, or sales promotional use. For information, please e-mail the Special Markets Department at SPsales@harpercollins.com.

A hardcover edition of this book was published in 2013 by Broadside Books, an imprint of HarperCollins Publishers.

FIRST BROADSIDE BOOKS PAPERBACK EDITION PUBLISHED 2014

Designed by Renato Stanisic

Library of Congress Cataloging-in-Publication Data has been applied for.

ISBN: 978-0-06-223174-1 (pbk.)

14 15 16 17 18   OV/RRD   10 9 8 7 6 5 4 3 2 1

FOR ANNABEL AND ALLEGRA,
WHO MADE ME PUT IN THE JENNY GEDDES STORY

CONTENTS

Acknowledgments

I t was only several years after being elected to political office that I finally admitted to myself that I would never be a full-time historian. There were too many more immediate challenges.

This book owes everything to the people who made a different choice, elevating the important over the immediate.

My thanks go, first, to Dr. Graham Stewart, one of Britain's finest modern historians, who offered priceless suggestions and improvements. Second, to Dr. James Hannam, who was kind enough to ameliorate great chunks of text. Third, to my tutor, Dr. Jeremy Catto: I can hardly, after his involvement with this book, call him my former tutor, though more than twenty years have passed since our last formal tutorial at Oxford.

Nor can I forget the other men who showed me how to look at the past: Richard Speight, Niall Hamilton, Michael Preston, Peregrine Horden, Richard Wilkinson, Robert Beddard, and Norman Stone.

Then there are the professional historians and writers whose thoughts have led mine, and whose insights show in this work as the shadows on the cave wall: James Bennett, Hugh Brogan, James

Campbell, Linda Colley, Dinesh D'Souza, Niall Ferguson, John Fonte, David Hackett Fischer, Lawrence James, Roger Kimball, Alan Maddicott, Alan Macfarlane, John O'Sullivan, Kevin Phillips, Andrew Roberts, Roger Scruton, Matthew Spalding, Claudio Véliz. Thanks, most of all, to Sara.

INVENTING FREEDOM

Introduction: The Anglosphere Miracle

There are few words which are used more loosely than the word "Civilization." What does it mean? It means a society based upon the opinion of civilians. It means that violence, the rule of warriors and despotic chiefs, the conditions of camps and warfare, of riot and tyranny, give place to parliaments where laws are made, and independent courts of justice in which over long periods those laws are maintained. That is Civilization—and in its soil grow continually freedom, comfort, and culture. When Civilization reigns, in any country, a wider and less harassed life is afforded to the masses of the people. The traditions of the past are cherished, and the inheritance bequeathed to us by former wise or valiant men becomes a rich estate to be enjoyed and used by all.

—WINSTON CHURCHILL, 1938

The liberty, the unalienable, indefeasible rights of men, the honor and dignity of human nature, the grandeur and glory of the public, and the universal happiness of individuals, were never so skillfully and successfully consulted as in that most excellent monument of human art, the common law of England.

—JOHN ADAMS, 1763

When I was four years old, a mob attacked our family farm. There was a back entrance, a footpath into the hills, and my mother led me there by the hand. "We're going to play a game," she told me. "If we have to come this way again, we must do it without making a sound."

My father was having none of it. He had a duty to the farmworkers, he said, and wasn't going to be driven off his own land by hooligans bused in from the city.

He was suffering, I remember, from one of those diseases that periodically afflict white men in the tropics, and he sat in his dressing gown, loading his revolver with paper-thin hands.

This was the Peru of General Juan Velasco, whose putsch in 1968 had thrown the country into a state of squalor from which it has only recently recovered. Having nationalized the main industries, Velasco decreed a program of land reform under which farms were broken up and given to his military cronies.

As invariably happens when governments plunder their citizens, groups of agitators decided to take the law into their own hands. It was the same story as in the Spanish Second Republic, or Allende's Chile: the police, seeing which way the wind was blowing, were reluctant to protect property.

Knowing that no help would come from the authorities, my father and two security guards dispersed the gang with shots as the latter attempted to burn down the front gates. The danger passed.

Not everyone was so lucky. There were land invasions and confiscations all over the country. The mines and fishing fleets were seized. Foreign investment fled and companies repatriated their employees. The large Anglo-Peruvian community into which I had been born all but disappeared.

Only many years later did it strike me that no one had been especially surprised. There was a weary acceptance that, in South America, property was insecure, the rule of law fragile, and civil government contingent. What you owned might at any moment be snatched away, either with or without official sanction. Regimes came and went, and constitutions were ephemeral.

At the same time it was assumed, by South Americans as well as by expatriates, that such things didn't happen in the English-speaking world. As I grew up, attending boarding school in the United Kingdom but returning to Peru for most of my vacations, I began to wonder at the contrast.

Peru, after all, was on paper a Western country. Its civilization was Christian. Its founders had thought of themselves as children of the Enlightenment, and had been strongly committed to reason, science, democracy, and civil rights.

Yet Peru—indeed, Latin America in general—never achieved the law-based civil society that North America takes for granted. Settled at around the same time, the two great landmasses of the New World serve almost as a controlled experiment. The north was settled by English-speakers, who took with them a belief in property rights, personal liberty, and representative government. The south was settled by Iberians who replicated the vast estates and quasi-feudal society of their home provinces. Despite being the poorer continent in natural resources, North America became the most desirable living space on the planet, attracting hundreds of millions of people with the promise of freedom. South America, by contrast, remained closer to the state of nature that the great philosopher Thomas Hobbes saw as the terrifying prelude to civil government. Legitimacy was never far removed from raw physical power, whether in the form of control of the mob or control of the armed forces.

It is hard to avoid the conclusion that this distinction reflects a difference between the two ancestral cultures. Please don't get me wrong. I am a convinced Hispanophile. I love Spanish literature and history, theater and music. I have spent happy times in every

Ibero-American state, as well as in sixteen of Spain's seventeen regions. I like the Hispanosphere precisely as it is. It's simply that, the more I have traveled there, physically and intellectually, the harder it is to sustain the idea that the English- and Spanish-speaking worlds are two manifestations of a common Western civilization.

What, after all, do we mean by Western civilization? What was Churchill driving at in his definition, quoted above? There are three irreducible elements. First, the rule of law. The government of the day doesn't get to set the rules. Those rules exist on a higher plane, and are interpreted by independent magistrates. The law, in other words, is not an instrument of state control, but a mechanism open to any individual seeking redress.

Second, personal liberty: freedom to say what you like, to assemble in any configuration you choose with your fellow citizens, to buy and sell without hindrance, to dispose as you wish of your assets, to work for whom you please, and, conversely, to hire and fire as you will.

Third, representative government. Laws should not be passed, nor taxes levied, except by elected legislators who are answerable to the rest of us.

Now ask yourself how many countries that are habitually labeled Western have consistently applied those ideals over, say, the past century. How many have an unshakable commitment to them even today?

That question began to nag at me insistently after I was elected to the European Parliament in 1999. The European Union is based on the premise that its twenty-eight member states share a common civilization. While their cultures might diverge at the margins, the theory goes, all sign up to the shared liberal democratic values of the West.

The reality is different. The three precepts that define Western civilization—the rule of law, democratic government, and individual liberty—are not equally valued across Europe. When they act collectively, the member states of the EU are quite ready to subordinate all three to political imperatives.

The rule of law is regularly set aside when it stands in the way of what Brussels elites want. To cite only the most recent example, the euro-zone bailouts were patently illegal. Article 125 of the EU Treaty is unequivocal: "The Union shall not be liable for, or assume the commitments of, central governments, regional, local or other public authorities, other bodies governed by public law, or public undertakings of any Member State." This clause was no mere technicality. It was on the basis of its promise that the Germans agreed to abandon the Deutschmark in the first place. As Angela Merkel put it: "We have a Treaty under which there is no possibility of paying to bail out states."

Yet, as soon as it became clear that the euro wouldn't survive without cash transfusions, the dots and commas of the treaties were set aside. Christine Lagarde, then the French finance minister and now the director of the International Monetary Fund, boasted about what had happened: "We violated all the rules because we wanted to close ranks and really rescue the euro zone. The Treaty of Lisbon was very straightforward. No bailouts."

To British eyes, the whole process seemed bizarre. Rules had been drawn up in the clearest language that lawyers could devise. Yet, the moment they became inconvenient, they were ignored. When the English-language press said so, though, it was mocked for its insular, Anglo-Saxon literal-mindedness. Everyone else could see that, as a Portuguese member of the European Parliament put it to me, "the facts matter more than the legislation."

Democracy, too, is regarded as a means to an end—desirable enough, but only up to a point. The European Constitution, later renamed the Lisbon Treaty, was repeatedly rejected in national referendums: by 55 percent of French voters and 62 percent of Dutch voters in 2005, and by 53 percent of Irish voters in 2008. The EU's response was to swat the results aside and impose the treaty anyway. Again, to complain was simply to demonstrate that English-speakers didn't understand Europe.

As for the idea that the individual should be as free as possible from state coercion, this is regarded as the ultimate Anglophone fetish. Whenever the EU extends its jurisdiction into a new field—decreeing what vitamins we can buy, how much capital banks must hold, what hours we may work, how herbal remedies are to be regulated—I ask what specific problem the new rules are needed to solve. The response is always the same: "But the old system was unregulated!" The idea that absence of regulation might be a natural state of affairs is seen as preposterous. In Continental usage, "unregulated" and "illegal" are much closer concepts than in places where lawmaking happens in English.

These places are generally lumped together, in Euro-speak, as "the Anglo-Saxon world." The appellation is not ethnic, but cultural. When the French talk of "les anglo-saxons" or the Spanish of "los anglosajones," they don't mean descendants of Cerdic and Oswine and Athelstan. They mean people who speak English and believe in small government, whether in San Francisco, Sligo, or Singapore.

It may come as a surprise to some American readers to learn that, in the eyes of many Continental European commentators, they and the British and the Australians and others form part of a continuous "Anglo-Saxon" civilization, whose chief characteristic is a commitment to free markets. American friends, in my experience, often bracket the United Kingdom with the rest of Europe, and emphasize the exceptionalism of their own story. Yet, as we shall see, very few foreigners think of Americans that way. Alexis de Tocqueville, who visited the United States in the early 1830s, is often quoted as a witness to that country's uniqueness. Quoted, but evidently not widely read, since on the very first page of *Democracy in America* he anticipates one of that book's main themes, namely the idea that

English-speakers carried a unique political culture with them to the New World and developed it there in ways far removed from what happened in French and Spanish America. "The American," he wrote, "is the Englishman left to himself."

Three times in the past hundred years, the free world has defended its values in global conflicts. In the two world wars and in the Cold War, countries that elevated the individual over the state contended against countries that did the opposite. How many nations were consistently on the side of liberty in those three conflicts? The list is a short one, but it includes most of the English-speaking democracies.

You might argue, of course, that this lineup simply reflects ethnic and linguistic kinship. Because the United Kingdom was at war, English-speakers around the world sympathized with the mother country. This is undeniably part of the explanation. I still become emotional when I recall the words spoken from his hospital bed by New Zealand's Labour prime minister, Michael Joseph Savage, a few hours after Britain's declaration of war on September 3, 1939: "With gratitude for the past and confidence in the future, we range ourselves without fear beside Britain. Where she goes, we go. Where she stands, we stand." Yet this is not the whole story. Look at the size of the war memorials outside Europe. Consider the sheer number of volunteers. During the Second World War, 215,000 men served from New Zealand, 410,000 from South Africa, 995,000 from Australia, 1,060,000 from Canada, 2,400,000 from India. The vast majority had made an individual decision to enlist.

What force pulled those young men, as it had pulled their fathers, halfway around the world to fight for a country on which, in most cases, they had never set eyes? Was it simply an affinity of blood and speech? Were the two world wars nothing more than racial conflicts, larger versions of the breakup of Yugoslavia or the Hutu-Tutsi massacres?

Not according to the governments who called for volunteers, nor to the men who answered that call. Soldiers are rarely given to sentimentality, but in the diaries and letters of the men who served in uniform, we find a clear conviction that they were defending a way of life—a better way than the enemy's. In both world wars, they believed that they were, in the slogan of the time, fighting for freedom.

Here is the radical newspaper the *West Indian* in 1915: "West Indians, most of whom are descendants of slaves, are fighting for human liberty together with the immediate sons of the Motherland."

Here is Havildar Hirram Singh writing to his family in India from the sodden trenches of northern France in the same year: "We must honor him who gives us our salt. Our dear government's rule is very good and gracious."

Here is a Maori leader in 1918, recalling the fate of native peoples in German colonies:

> We know of the Samoans, our kin. We know of the Eastern and Western natives of German Africa. We know of the extermination of the Hereros, and that is enough for us. For 78 years we have been, not under the rule of the British, but taking part in the ruling of ourselves, and we know by experience that the foundations of British sovereignty are based upon the eternal principles of liberty, equity and justice.

We can easily slip into thinking that the values now prevalent in the world, the values we call Western, were somehow bound to triumph in the end. But there was nothing inevitable about their victory. Had the Second World War ended differently, liberty might have been beaten back to North America. Had the Cold War gone the other way, it could have been extirpated altogether. The triumph of the West was, in practical terms, a series of military successes by the English-speaking peoples.

It is, of course, undiplomatic to say so, which is why writers and politicians are so much more comfortable using the term *Western* than *Anglosphere*. But what do we mean by Western? During the Second World War, the designation was used to mean the countries attacking Nazi Germany from that direction. Through the long decades of the Cold War, it meant members of NATO and their allies on other continents.

After the fall of the Berlin Wall, a new definition quickly became current. In a lecture in 1992, later turned into an essay and then a book, Samuel Huntington divided the world into broad cultural spheres. He entitled his thesis *The Clash of Civilizations* and forecast (incorrectly, so far) that conflicts would increasingly take place between rather than within these spheres. Huntington looked for the origins of the West in the division between Latin and Greek branches of the Christian Church, a division that became a formal schism in 1054. The West, by this demarcation, is made up of those European nations that are predominantly Catholic or Protestant rather than Orthodox in their culture, plus the United States, Canada, Australia, and New Zealand.

Such a definition correlates fairly closely with Western military structures. Yet these structures, in their present form, are recent. Several countries now in NATO were, within living memory, allied to Hitler or Stalin or both. Indeed, outside the Anglophone world, the list of states with more or less continuous histories of representative government and freedom under the law is shorter than anyone likes to admit: Switzerland, the Netherlands, the Nordic countries.

As Mark Steyn has put it, penetratingly if indelicately:

Continental Europe has given us plenty of nice paintings and agreeable symphonies, French wine and Italian actresses and whatnot, but, for all our fetishization of multiculturalism, you can't help noticing that when it comes to the notion of a political West—one with a sustained commitment to liberty and democracy—the historical record looks a lot more unicultural and, indeed (given that most of these liberal democracies other than America share the same head of state), uniregal. The entire political class of Portugal, Spain, and Greece spent their childhoods living under dictatorships. So did Jacques Chirac and Angela Merkel. We forget how rare on this earth is peaceful constitutional evolution, and rarer still outside the Anglosphere.

Ideological borders move more swiftly than physical ones. A wave of European states embraced Western values after 1945, and another wave after 1989. But when we use "Western values" in this context, we're being polite. What we really mean is that these countries have adopted the characteristic features of the Anglo-American political system.

Elected parliaments, habeas corpus, free contract, equality before the law, open markets, an unrestricted press, the right to proselytize for any religion, jury trials: these things are not somehow the natural condition of an advanced society. They are specific products of a political ideology developed in the language in which you are reading these words. The fact that those ideas, and that language, have become so widespread can make us lose sight of how exceptional they were in origin.

Let me make a sartorial analogy. H. G. Wells once observed that the English were unique among the nations of the world in having no national dress. He was wrong—and wrong in a telling way. The national dress of the English—a suit and tie—has ceased to seem English, because it is worn all over the planet. On formal occasions, men in most countries dress as Englishmen; the rest of the time they dress, for the most part, as Americans, in jeans.

There are a few redoubts, of course. You occasionally see Bavarian men in leather shorts alongside women in dirndls. Some Arabs have kept their robes and headscarves. But, by and large, the Anglosphere has lost its distinctive apparel. Such was the power of the industrial revolution—which was, before anything, a revolution in textiles—that, during the twentieth century, the English-speaking peoples clothed the world in their image—and, in doing so, forgot that the global costume was really theirs.

It is natural, when we think of a country, to focus on the things that make it different rather than the things that it has exported successfully. When people are asked to name a British food, they will be likelier to say "steak-and-kidney pie" than "a sandwich." When asked to name an English sport, they will pick cricket rather than football. And so it is with values. Asked what the identifying features of the British political system are, foreigners and Britons alike will often point to the monarchy, the House of Lords, the maces and horsehair wigs and other trappings of parliamentary procedure. Asked the same question about the United States, they will talk of the exorbitant cost of campaigns, the insidious corporate donations, the vicious attack ads. In neither case are they likely to identify the truly extraordinary feature, namely that the lawmakers are answerable to everyone else, and that governments change peacefully as a result of popular votes.

The rule of law is rarer than we sometimes realize. Oppression and arbitrary power are far more usual. Man is a competitive creature, domineering and rapacious when the circumstances are right. Politically, a medieval European monarchy would not have been so very different to a modern African kleptocracy. Once people are in a position to set the rules, they tend to rig those rules in their own favor. Obedient to the promptings of their genome, they design the system so that their descendants, too, will enjoy an advantage over everyone else. Arbitrary power, hereditary status, the systematic looting of resources by the ruling caste: these things were once near universal and are still the norm for most human beings. The real question is not whether liberal democracy was always destined to succeed, but how it managed to get off the ground at all.

We are still experiencing the aftereffects of an astonishing event. The inhabitants of a damp island at the western tip of the Eurasian landmass stumbled upon the idea that the government ought to be subject to the law, not the other way around. The rule of law created security of property and contract, which in turn led to industrialization and modern capitalism. For the first time in the history of the species, a system grew up that, on the whole, rewarded production better than predation. That system proved to be highly adaptable. It was taken across the oceans by English-speakers, sometimes imposed by colonial administrators, sometimes carried by patriotic settlers. In the old courthouse in Philadelphia, it was distilled into its purest and most sublime form as the U.S. Constitution.

So successful was the model that almost every state in the world now copies at least its trappings. Even the most brutal dictatorships generally have things called congresses, whose nervous delegates, anticipating the wishes of the autocrat, group themselves into blocs called political parties. Even the nastiest despotisms have institutions called supreme courts, which, on paper, are something other than an instrument of the regime. But meaningful political freedom— freedom under the rule of law in a representative democracy— remains an unusual phenomenon. We make a mistake when we assume that it will necessarily outlast the hegemony of the English-speaking peoples.

This book tells the story of freedom—which is to say, it tells the story of the Anglosphere. I realize that this statement might strike some readers as smug, triumphalist, even racist. But I hope, over the course of the story, to show you that it is none of those things. From the first, the Anglosphere was a civil rather than an ethnic concept: that was a large part of its strength. While a few Victorian writers tried to explain the success of the English-speaking peoples in racial terms, their arguments were controversial even at the time and are untenable today. The reason that a child of Greek parents in Melbourne is wealthier and freer than his cousin in Mytilene has nothing to do with race and everything to do with political structures.

Part of the problem lies with the vagueness of the terminology. *Anglosphere* is a word of recent coinage, first used in a Neal Stephenson's 1995 science fiction novel, *The Diamond Age*. It spread rapidly into our political and cultural vocabulary because it described something for which a word was needed, namely the community of free English-speaking nations. *The Oxford English Dictionary* defines Anglosphere as "the group of countries where English is the main native language," but the man who popularized the concept, the American writer James C. Bennett, is more exacting in his criteria:

> To be part of the Anglosphere requires adherence to the fundamental customs and values that form the core of English-speaking cultures. These include individualism, the rule of law, honoring contracts and covenants, and the elevation of freedom to the first rank of political and cultural values. Nations comprising the Anglosphere share a common historical narrative in which the Magna Carta, the English and American Bills of Rights, and such Common Law principles as trial by jury, presumption of innocence and "a man's home is his castle" are taken for granted.

Which nations? All definitions include five core countries: Australia, Canada, New Zealand, the United Kingdom, and the United States. Almost all count Ireland, whose special circumstances are discussed later. Most also take in Singapore, Hong Kong, and what's left of Britain's colonial archipelago (Bermuda, the Falkland Islands, and so on). Some also encompass the more democratic Caribbean states, and some embrace South Africa.

The elephant—for once the metaphor seems entirely apt—is

India, which, if included, would constitute two-thirds of the Anglo-sphere's population. India, too, is discussed separately later on.

It was once uncontroversial to see the spread of liberty as being bound up with the rise of the Anglosphere. After the Reformation, many English-speakers saw the ascendancy of their civilization as providential. Theirs was the new Israel, a chosen nation, appointed by God to carry freedom across the world. The opening lines of "Rule, Britannia," that hymn to British liberty, are so often belted out that we rarely stop to listen: "When Britain first *at Heaven's command* arose from out the azure main . . ." The same conviction, in an even more intense form, motivated the first Americans.

The religious impulse faded with the years, but the belief in destiny did not. British and American historians pointed to a series of events that had brought their ancestors ineluctably toward modernity and greatness: the establishment of the common law, Magna Carta, the Grand Remonstrance, the Bill of Rights, the U.S. Constitution, the scientific revolution, the abolition of slavery.

During the twentieth century, such flag-waving views of the past became unfashionable. As Marxism, anticolonialism, and multiculturalism came into vogue, historiography altered. The writers who had celebrated the great political milestones of Anglo-American history were charged with having been complacent, culturally arrogant, and, worst of all, anachronistic.

Their error, it was said, had lain in seeing a pattern in events that would have been invisible to participants. The patriotic historians, argued critics, had tended to see major historical crises as steps toward the apex of human civilization—a golden age that they conveniently situated in their own lifetimes.

In 1931, a Cambridge professor, Herbert Butterfield, published *The Whig Interpretation of History*, perhaps the single most influential work of historiography ever written. Whig historians, he argued, made the mistake of seeing past events teleologically—that is, as movements toward a fixed destination. In fact, the actors in those events often had very different motives to those of their modern cheerleaders. Teleological history led writers into the folly of dividing historical figures into good guys (those who supported Whig and liberal policies, such as a wider franchise and the spread of civil rights) and bad guys (those who resisted progress). As Butterfield put it, "The study of the past with one eye upon the present is the source of all sins and sophistries in history. It is the essence of what we mean by the word 'unhistorical.'"

Many of Butterfield's criticisms hit home, and his book changed forever the way in which history was written in English. Historians began to grasp, for example, that the opponents of royal power were often, in their own eyes, not progressives but conservatives, defending what they believed to be an ancient constitution against a modernizing court.

Yet the weaknesses of Whig history should not detract from its verities. The events that the Whig historians freighted with importance—Magna Carta, the Reformation, the Petition of Right, the Philadelphia Constitutional Convention—were seen in much the same light by contemporaries. And while it may be anachronistic to label dead men as good or bad on the basis of how closely their views resemble a later generation's, it is also impossible to write meaningful history without value judgments.

The Whig historians glimpsed important truths. Modern research tends to sustain their view that constitutional liberty has its roots in pre-Norman England. The exceptionalism they took for granted, and from which most twentieth-century historians flinched for fear of being thought supremacist or racist, turns out to be real enough. It is even possible to discern, as they did, two enduring factions within the English-speaking peoples: one committed to the values that underpinned that exceptionalism, and one hankering after the more statist models favored in the rest of the world. To label these factions "Whig" and "Tory" is, without question, anachronistic; yet it is also an invaluable shorthand.

The categorization, after all, was not an invention of the Whig historians. It was understood by many of the key agents of the events they described. Thomas Jefferson explained it in characteristically partisan terms:

> The division into Whig and Tory is founded in the nature of man; the weakly and nerveless, the rich and the corrupt, seeing more safety and accessibility in a strong executive; the healthy, firm, and virtuous, feeling confidence in their physical and moral resources, and willing to part with only so much power as is necessary for their good government; and, therefore, to retain the rest in the hands of the many, the division will substantially be into Whig and Tory.

Being a Whig, for Jefferson and his followers, didn't simply mean a general affinity with manliness, independence, and republican virtue. It was a specific identification with an ancient English cause. One popular pamphlet published in 1775 defined the Patriots' creed as resting on "the principles of Whigs before the Revolution [the Glorious Revolution of 1688] and at the time of it."

What were these principles? The pamphleteer listed them concisely. Lawmakers should be directly accountable through the ballot box; the executive should be controlled by the legislature; taxes should not be levied, nor laws passed, without popular consent; the individual should be free from arbitrary punishment or confiscation; decisions should be taken as closely as possible to the people they affected; power should be dispersed; no one, not even the head of state, should be above the law; property rights should be secure; disputes should be arbitrated by independent magistrates; freedom of speech, religion, and assembly should be guaranteed.

There is a reason that supporters of these precepts, both in Britain and in America, called themselves "patriots." They could see something that later generations affected not to see: that the liberties they valued were largely confined to the English-speaking world; and that their domestic opponents wanted to bring their political system into line with more autocratic foreign models.

The tragedy of our age is that those domestic opponents are succeeding. Having developed and exported the most successful system of government known to the human race, the English-speaking peoples are tiptoeing away from their own creation.

Britain's intellectual elites see Anglosphere values as an impediment to assimilation into a European polity. Their equivalents in Australia see them as a distraction from their country's supposed Asian destiny. In the United States, especially under the present administration, Anglosphere identity is seen as a colonial hangover, the patrimony of dead white European males. In every English-speaking country, a multiculturalist establishment hangs back from teaching children that they are heirs to a unique political heritage.

Consequently, in most Anglosphere states, the "principles of the Whigs before the Revolution" are being slowly abandoned. Laws are now regularly made without parliamentary approval, taking the form of executive decrees. Taxes are levied without popular consent, as during the bank bailouts. Power is shifting from local, provincial, or state level to national capitals, and from elected representatives to standing bureaucracies. State spending has grown to a level that earlier Anglosphere populations would have regarded as a cause for popular revolt. If we want to understand why the Anglosphere hegemony is failing, we need look no further.

The owl of Minerva, wrote Hegel, spreads its wings only with the gathering of the dusk. As the sun sets on the Anglosphere imperium, we understand with sudden clarity what it is that we stand to lose. What raised the English-speaking peoples to greatness was not a magical property in their DNA, nor a special richness in their earth, nor yet an advantage in military technology, but their political and legal institutions.

The happiness of the human race depends, more than anyone likes to admit, on the survival and success of those institutions. As a devolved network of allied nations, the Anglosphere might yet exert its benign pull on the rest of this century. Without that pull, the future looks altogether grayer and colder.

CHAPTER 1

The Same Language, the Same Hymns, the Same Ideals

We must be free or die, who speak the tongue
That Shakespeare spake, the faith and morals hold
Which Milton held.

—WILLIAM WORDSWORTH, 1807

The colonies planted by England in America have immeasurably out-grown in power those planted by Spain. Yet we have no reason to be-lieve that, at the beginning of the sixteenth century, the Castilian was in any respect inferior to the Englishman. Our firm belief is, that the North owes its great civilization and prosperity chiefly to the moral effect of the Protestant Reformation, and that the decay of the southern countries of Europe is to be mainly ascribed to the great Catholic revival.

—LORD MACAULAY, 1840

A GREAT HOUR TO LIVE

It was the longest walk of his presidency. For twenty years, Frank-lin D. Roosevelt had hidden his polio from the American public:

photographs showed him standing unaided, or sitting in chairs. This time, though, invited by Winston Churchill to join him on HMS *Prince of Wales*, he insisted on walking. His aides had tried to talk him out of it: what if the deck were to pitch, leaving the president in an undignified heap? But FDR was determined to rise, literally, to the occasion. Supported by his son on one side and a naval officer on the other, and leaning heavily on his cane, he made his slow way to where the British prime minister waited, while the band of the *Prince of Wales* played "The Stars and Stripes Forever."

The date was August 10, 1941, and what followed was a carefully staged and brilliant display of what bound the two great English-speaking powers together. The location was Placentia Bay, off Newfoundland. Churchill thought of Canada rather as he thought of himself, as a living embodiment of Anglosphere unity. As he was to tell MPs in Ottawa a few months later, "Canada occupies a unique position in the British Empire because of its unbreakable ties with Britain and its ever-growing friendship and intimate association with the United States."

The U.S. president had been edging his country away from the neutrality that its founders had decreed, and which had been reinforced by a series of laws passed in the 1930s. While he could not enter the war, Roosevelt was determined to support the rest of the Anglophone world.

The United States, FDR had told his countrymen, should be "the arsenal of democracy." He had made warships and matériel available to the United Kingdom in exchange for a ninety-nine-year lease on various military bases, thereby tilting the scales, if only slightly, back toward Britain. He had decreed that the Royal Navy might use U.S. facilities, and initiated a formal collaboration between the two air forces. Most recently, he had made substantial resources available to the British government on easy credit terms: the so-called Lend-Lease Agreement.

These accords had been popular on both sides of the Atlantic. While a handful of Tories complained about the establishment of sovereign U.S. bases within the empire, most Britons accepted that the interests of the English-speaking peoples were now permanently aligned, as indeed they were to prove: when, during the campaign against the Taliban in 2001, the U.S. flew missions from one such base, in the British Indian Ocean Territory, no one in either country raised an eyebrow, so natural was the partnership. American voters, for their part, approved of Lend-Lease because they overwhelmingly hoped for a British victory.

To a certain extent, of course, theirs was the sympathy of one civilian democracy for another. Germany and its satellites were fascist dictatorships; Denmark, Norway, France, Belgium, and the Netherlands, all of which had been occupied by the Nazis, had had free parliaments. The political similarities between Britain and America, though, rested upon a deeper affinity, and it was this affinity that Churchill was determined to display.

It was a Sunday morning, and the crew of the USS *Augusta*, which had carried Roosevelt to Newfoundland, was paraded alongside that of HMS *Prince of Wales* for a religious service. Churchill had been determined that "every detail be perfect," and the readings and hymns were meticulously chosen. The sailors listened as a chaplain read from Joshua 1 in the language of the King James Bible, revered in both nations: "As I was with Moses, so will I be with thee: I will not fail thee, nor forsake thee. Be strong and of good courage." The liturgy was equally familiar to the British and American seamen. It was, remarked Churchill's private secretary, "a sort of marriage service."

The prime minister himself was delighted. "The same language, the same hymns and, more or less, the same ideals," he enthused, later adding: "Every word seemed to stir the heart, and none who took part in it will ever forget the spectacle presented that sunlit morning on the crowded quarterdeck—the symbolism of the Union Jack and the Stars and Stripes draped side by side on the pulpit. It was a great hour to live."

Wars have a way of making kindred peoples forget their differences. The United Kingdom and the United States had not quarreled since 1895, when the Cleveland administration intervened in a border dispute between Venezuela and British Guiana. In reality, there had been no serious prospect of war between the two states since the 1820s. The Monroe Doctrine, whereby the United States rejected the interference of any European power in its hemisphere, had rested on the support of the Royal Navy, and both governments knew it.

Britain had tacitly backed the United States in its 1898 war against Spain, and Presidents McKinley and Roosevelt had returned the favor during the Boer War, despite opposition from Irish-American voters. The two countries had been allies in the First World War: the Kaiser's government had taken it for granted that the "Anglo-Saxon" powers would stick together, and had invited Mexico to invade its northern neighbor, thereby prompting U.S. entry into the war.

Though neither Churchill nor Roosevelt yet knew it, Hitler was to make the same assumption: there are few other ways to explain his decision to declare war on the United States in the aftermath of Pearl Harbor. The differences between the Anglosphere nations, so important to their peoples, have always been less apparent to outsiders.

Churchill, whose mother was American, and who had spent three years working on his monumental *History of the English-Speaking Peoples* before the war interrupted (the four volumes were eventually published in the late 1950s), understood better than anyone else that the Atlantic Alliance couldn't rest simply on coincident interests; it had to be based on a sense of shared identity.

Three key elements of that shared identity were on display that Sunday morning off Newfoundland. They were the very elements listed by Churchill: "the same language, the same hymns and, more or less, the same ideals." Let us consider them in turn.

THE SAME LANGUAGE

Language is the usual denominator of nationality. It's true that there are others: nations have sometimes self-defined on the basis of history, geography, or religion. Language, however, is a more frequent determinant than these, for it is a prerequisite of mutual understanding. As Rudyard Kipling, an Anglospherist *avant la lettre*, put it:

> *The Stranger within my gate,*
> *He may be true or kind,*
> *But he does not talk my talk—*
> *I cannot feel his mind*
> *I see the face and the eyes and mouth*
> *But not the soul behind.*

For most of their history, the English-speaking peoples have been dispersed across different states. Only twice has a single polity encompassed more or less the entire language group. The first occasion was under Cromwell's imperium, between 1653 and 1660, and it was imposed by force. The second was between 1707 and 1776, less than the span of a single human lifetime. The fracturing of that polity did not turn its people overnight into mutual foreigners. When John Adams, the future president, presented his credentials to George III as the first ambassador of an independent American republic, he spoke some touching words that moved his former enemy almost to tears:

> I shall esteem myself the happiest of men if I can be instrumen-
> tal in restoring the confidence and affection or, in better words,
> the old good nature and old good humor, between peoples who,
> though separated by an ocean and under different governments,
> have nonetheless the same language, a similar religion and kin-
> dred blood.

A shared language makes possible the jokes, the nuances, the subtleties that flesh out human relations. These things, in turn, create fellow-feeling. During the Boer War, a former Rough Rider wrote from South Africa to his old commander, Theodore Roosevelt, now vice president of the United States: "Dear Teddy, I came over here meaning to join the Boers, who I was told were Republicans fighting Monarchists; but when I got here I found the Boers talked Dutch while the British talked English, so I joined the latter."

Language creates a sense of sympathy that extends beyond state borders. During the Falklands War, which coincided with the 1982 Football World Cup in Spain, Spanish crowds chanted "Argentina! Argentina!" at every England match. The United States, by contrast, was in no doubt about where it stood: a motion condemning the invasion and calling for an Argentine withdrawal passed the Senate with only one vote against (that of the contrarian Jesse Helms).

When, in 1995, Spain was involved in a fishing dispute with Canada, towns across the United Kingdom, and especially in fishing communities, flew maple leaf flags. Although the EU as a whole lined up with Spain on that occasion—Germany went so far as to send naval vessels to the area—its two Anglophone members, Ireland and the United Kingdom, backed Canada.

In both cases, language trumped geography. People didn't choose sides on the basis of whether they were in Europe or the Americas, in the EU or the Organization of American States, in the Old World or the New. Nor, for the most part, was their sympathy determined by the strict merits of the dispute. Instead, unsurprisingly, they backed the folk of their own speech against those they couldn't understand.

What is perhaps more surprising is a property that may be particular to the language in which I'm writing. English doesn't simply unite its speakers in the sense that it lets them read the same books, watch the same television programs, and sing the same songs. It also seems to have inherent qualities that facilitate a certain view of the world.

Anyone who speaks more than one language knows that when you switch from thinking in one to thinking in another, your perspective slightly adjusts. Words don't always have precise equivalents, and some ideas become marginally altered in translation. Even if a sentence is rendered in the same way, word for word, the connotations can be different.

For example, "democracy" has the same literal meaning in almost every language: a system where decisions are made by majority vote. But in English, the overtones are so positive that *democracy* is now used, more generally, as an all-purpose hurray-word. People will say things like "Private schools are undemocratic"—meaning, in effect, nothing more than "I disapprove of private schools." In Russian or Chinese, the word for democracy has no such associations. It simply denotes one among several competing theories of how to organize society.

There is an intriguing connection between the global dominance of English and the ideas that were first conceived and expressed in that language. As the philologist Robert Claiborne puts it, "The tongue and the philosophy are not unrelated. Both reflect the ingrained Anglo-American distrust of unlimited authority, whether in language or in life." Madhav Das Nalapat, who holds the UNESCO Peace Chair and is the director of the Department of Geopolitics at Manipal University, argues that "the spread of education in English has reduced the attraction of radicalism."

Could he be right? Could it be that the advance of democracy
and free trade over the past six decades is connected to the rise of
English as the first planetary language?

Our speech, uniquely, spilled out from behind its imperial
frontiers. English was by no means the first tongue to have been
disseminated through migration and conquest: Aramaic, Greek,
Latin, Persian, Arabic, Russian, Dutch, French, and Spanish were
all imperial idioms. But consider what happened to them after de-
colonization. Most endured only where there was a large population
of native speakers. Young Indonesians are now far likelier to speak
English than Dutch. Spanish is secure enough in Central and South
America but has been almost wholly ousted from the Philippines.
Only elderly Syrians or Vietnamese are apt to speak decent French.
Indeed, the displacement of French by English has an unmissable
political dimension. When, for example, the *génocidaire* regime was
toppled in Rwanda, the new government ordered schools to teach
English rather than French, and applied successfully to join the
Commonwealth, intending with both measures to signal its com-
mitment to liberty.

In 1492, when English was still the vernacular of a rude archipelago, the Spanish scholar Antonio de Nebrija presented Queen
Isabella with a comprehensive grammar of the Castilian language.
Christopher Columbus had set sail a few weeks earlier, and it would
be many months before he returned with his world-changing news,
yet Nebrija had no doubt that Spain was on the way to imperial
greatness. In the dedication to his sovereign he included a phrase
whose timing, in the circumstances, was uncanny: "My certain conclusion is that language is the companion of empire."

In that sentence, there is a clue to why Spain's colonies in the
New World were to develop so differently from England's. It would
not have occurred to any English-speaker at that time to regulate his
language, in whose cadences could yet be heard the relatively recent

fusion of Anglo-Saxon and Norman-French. And what went for his language went for his government.

English-speaking America was dispersed and libertarian, containing neither an episcopacy nor an aristocracy, and reliant on the local self-government of the settlers. Spanish-speaking America was, from the start, a governmental project, initially run by the Crown and the Jesuits but, after 1787, by the Crown alone, the Jesuits having been accused of "building an empire within an empire" and suppressed—a reminder that, in every authoritarian system, power is jealous.

To the Spanish authorities, control of the language was every bit as natural as control of the governmental structure of Mexico and Peru. Such control, whether linguistic or political, stunts development. Constrained by statist thinking, Hispanic America never quite fulfilled its potential, which is why it has tended to export people to the United States rather than the reverse; and why, despite its enormous advantages in terms of population and territory, and despite being arguably the easiest mainstream European language for a nonspeaker to learn, Spanish never became a major international medium.

Consider, for a moment, the contrast. As the English language became globalized, so did a series of Anglosphere customs and institutions, from bicameral legislatures to the Boy Scouts, from stock exchanges to golf, from jury trials to horse races. I could fill the rest of the book with examples of successful Anglophone exports, but it is more instructive to ponder, as a contrast, the relative limits of Spanish cultural exports. Spain, after all, was the world's leading power when the New World was first settled. And yet, the Chilean historian Claudio Véliz observes:

> How difficult it is to discover cultural traits and artifacts of Spanish origin that succeeded in gaining universal acceptance. There are a few, some of them truly distinguished, like the Catholic

Reformation and all its institutional and doctrinal concomitants, but after the inclusion of a couple of glorious archetypes (Don Juan and Don Quixote) the Society of Jesus, the common canary and the word liberalism, Merino sheep and the modern design of the Persian guitar, the list thins rapidly: bullfighting and castanets traveled badly.

Véliz has a simple explanation for the contrast. Anglophone culture, he argues, was adaptable, Hispanophone culture rigid. In the seventh century BC, the Greek mercenary and poet Archilochus wrote: "The fox knows many tricks. The hedgehog knows one big one." The Anglosphere, for Véliz, is the fox to the Hispanosphere's hedgehog: Anglophone culture, like the English language, is decentralized and individualistic. Other empires rose and fell, their cultural impact dwindling as they declined. The vulpine Anglosphere and its language outlasted them all.

English is today the official language of almost every international association: ASEAN, NATO, the World Bank, the IMF, APEC, OPEC. It is commonly used even in organizations where none of the member states is Anglophone, such as the European Free Trade Association.

My own place of work, the European Parliament, used to be one of the last redoubts of French preeminence and, in all its formal dealings, still gives equal weight to its twenty-four languages. But, from the mid-1990s, and especially after the admission of the formerly communist states in 2004, English became its unofficial common tongue. The handful of members of the European Parliament who haven't mastered the language are at an immense disadvantage. While they can perform well enough in plenary and committee sessions, where simultaneous interpretation is provided, they cannot join in the conversations in the corridors where the deal-making takes place.

Working in that multilingual environment has convinced me that there are intrinsic properties in English that favor the expression of empirical, down-to-earth, practical ideas. I often listen to the interpretation with my headphones covering one ear, so as to improve my language skills. Frequently, a politician or official will say something that seems to make sense enough in his own tongue but that, when rendered into English, turns out to be so abstract as to be almost meaningless.

The Australian philosopher David Stove noticed the same phenomenon. In a remarkable essay called "What Is Wrong with Our Thoughts?" he examined a series of catastrophically badly phrased academic ideas. He found them, not by scraping the underbelly of Marxism or Freudianism, but by reading some of the better-regarded writers in the Western canon: Plotinus, Hegel, Foucault. He quoted their words in translation, explaining: "it is very striking that I *had* to go to translations. Nothing which was ever expressed originally in the English language resembles, except in the most distant way, the thought of Plotinus, or Hegel, or Foucault. I take this to be enormously to the credit of our language."

I'm afraid Stove was mistaken in one regard. Plenty of academic papers in English are now written in unintelligible cant, the authors evidently confusing opacity of expression with profundity of thought. But such authors generally also look to statist European thinkers when it comes to their view of how to organize society, which rather proves Stove's point.

The loveliest language in the world can fail to move the determinedly abstract writer. Karl Marx was every bit as stodgy and meaningless when discussing Shakespeare as when discussing economics. Here, for example, is what he wrote after watching *Timon of Athens*:

> Since money, as the existing and active concept of value, confounds and confuses all things, it is the general confounding and confusing of all things, the world upside down, the confounding and

confusing of all natural and human qualities. If money is the bond
binding me to human life, binding society to me, connecting me
with nature and man, is not money the bond of all bonds? Can it
not dissolve and bind all ties? Is it not, therefore, also the universal
agent of separation?

What native speaker of English could have written that way?
Only one who had trained himself, over many years, to ape the style
of Hegel or Marx, Derrida or Sartre.

The solidity and pragmatism of English has a great deal to do
with the way it evolved. English-speakers thought of their language
as they thought of their legal and political institutions, as the property of the people rather than of the state. Just as the common law
grew like a coral, case by case, without central supervision, so did
the English language.

Neighboring states established committees and academies to regulate how their languages should be written and spoken. Their most
illustrious citizens would pronounce on what constituted acceptable
grammar, syntax, and orthography. The Académie Française was
founded by Cardinal Richelieu in 1635, the Real Academia Española
(Royal Spanish Academy) by Philip V in 1714. To this day, it hard to
find a higher honor than appointment to such an institute. Yet their
effect, by definition, is to limit and circumscribe the development of
correct speech.

English, unusually, has no such restraints. The closest thing it
has to an arbitrator of correct vocabulary is, in the United States,
Merriam-Webster's dictionary (first published in 1828, by Noah
Webster), and, in the United Kingdom, *The Oxford English Dictionary* (first published 1928). Both are private undertakings.

English is protean, freewheeling, and voracious. It contains more
than twice as many words as French, more than three times as many
as Spanish. In part this variety reflects the fact that modern English

emerged from a multilingual society, in which the bulk of the population spoke Old English, the upper classes spoke Norman-French, and most writing was done in Latin. As a result, many words can take at least three forms, with alternative Anglo-Saxon, French, and Latin etymologies: *rise, mount, ascend*; *ask, question, interrogate*; *time, age, era*; *goodness, virtue, probity.*

But the bigger factor is that English was unregulated, and therefore free to assimilate anything it found useful. The most famous English dictionary was published by Samuel Johnson in 1755, three years of work by a single, brilliant, idiosyncratic mind. The first French dictionary, published in 1694, had been compiled over forty years by forty editors. "Forty times forty is sixteen hundred," declared the dyspeptic doctor. "As three to sixteen hundred, so is the proportion of an Englishman to a Frenchman." This idea so tickled Johnson's actor friend, David Garrick, that he composed a patriotic, if clumsy, couplet in celebration:

> *Johnson, well arm'd like a hero of yore*
> *Has beat forty French and will beat forty more.*

The sheer number of words in English allows the writer to express himself cleanly and unambiguously. If he instead chooses to write in the manner of Foucault or Gramsci, it is because he is being deliberately obscure.

George Orwell, one of the finest prose stylists of any age or nation, understood that the scarcity of words was a potential enemy to truth and freedom. In *Nineteen Eighty-Four,* he has the Newspeak philologist Symes explain how independent thinking can be prevented by the reduction of vocabulary:

> You think, I daresay, that our chief job is inventing new words. But
> not a bit of it! We're destroying words—scores of them, hundreds

of them, every day. We're cutting the language right down to the bone. Don't you see that the whole aim of Newspeak is to narrow the range of thought? In the end we shall make thoughtcrime literally impossible, because there will be no words in which to express it.

Conversely, a language with an expansive and unfettered vocabulary can be used to say almost anything. If no word exists to convey a useful idea, one can be invented—such as *Anglosphere*.

Is it fanciful to posit a direct connection between the English language and the distinctive political system of the Anglosphere? I don't think so. Most of the libertarian lexicon was originally English. Andrew Roberts identifies some early English coinages as "liberty of conscience" (1580), "civil liberty" (1644), and "liberty of the press" (1769). How much is correlation and how much causation? It is hard to say. Actual press freedom, for instance, has existed in English law since 1695—a century after the first recorded use of the phrase "liberty of conscience" and a century before "liberty of the press." But that there is some link is almost unarguable.

Think about the most famous apologia for democracy ever uttered. On November 19, 1863, at the Soldiers' National Cemetery in Gettysburg, Pennsylvania, President Abraham Lincoln, weak and lightheaded with an oncoming case of smallpox, made a speech that lasted for just over two minutes, and ended with his hope "that this nation, under God, shall have a new birth of freedom—and that government of the people, by the people, for the people, shall not perish from the earth."

Those words have been quoted ever since, as the supreme vindication of representative government. Indeed, they are often quoted as proof of American exceptionalism. But the words were not Lincoln's. Most of his hearers would have recognized their source, as our generation does not. They came from the prologue to what was probably the earliest translation of the Holy Scriptures into the English language: "This Bible is for the government of the people, for the people and by the people." The author was the theologian John Wycliffe, sometimes called "the Morning Star of the Reformation." Astonishingly, the words had first appeared in 1384.

In no other language could such a concept have been verbalized at that time. The English language has been both a vehicle and a guarantor of liberty down the centuries. As the two leaders sat down together on the deck of HMS *Prince of Wales* in August 1941, comfortable in that total understanding that can only exist between people speaking the same tongue, it was about to play that role again, to the immense benefit of mankind.

THE SAME HYMNS

John Wycliffe, the original author of Abraham Lincoln's famous words, is the most arresting figure in the medieval English church. Philosopher, temperamental rebel, and heresiarch, he anticipated many of the doctrines and practices of Protestantism. He believed that the church had amassed too much wealth and was drifting away from the example of Jesus. He opposed the selling of indulgences, rejected transubstantiation, and emphasized salvation by faith. He thought that priests should be allowed to marry, and that they should be accountable before the civil courts like everyone else. He rejected papal authority in England, arguing that the nation was bound instead to its own Crown and institutions.

Above all, and exceptionally for his time, Wycliffe believed in the centrality of the Bible. He taught that people should read the scriptures for themselves and not rely on the interpretation of priests and prelates. In his last years, he devoted himself to translating the Bible from Latin into English. The Holy See considered him such an abominable heretic that, forty-four years after he had died, his bones were dug up, ground to powder, and cast into a river.

"Government of the people, by the people, for the people" was, in Wycliffe's mind, a concept with political, religious, and educational implications. If men and women were free to make up their minds on theological questions, they would also be better suited to independence in secular affairs.

The association between religious and civil freedom was to become a central tenet of the Anglosphere, critical to the self-definition of its peoples. National history tended to be understood as the providential triumph of those values in the face of oppression—and not without reason. From the beginning, supporters of liberty, whether sacred or profane, had indeed been resisted by the beneficiaries of the existing feudal order.

In 1381, there was a major uprising by the poorer classes against the conditions of serfdom. In the crackdown that followed, Wycliffe's followers, who were known as Lollards, were rounded up along with the rebels.

For the next century and a half, until the arrival of Protestantism in England, Lollardy survived as a clandestine and largely lower-class movement, disseminating its doctrines in private homes. A Bible-based sect, it was limited by its lack of access to a printing press—the huge advantage enjoyed by the Continental reformers who reached England in the 1530s. As A. G. Dickens, the supreme authority on the English Reformation, put it, "Lollardy created an underground, and there awaited the appearance of liberators. When liberation finally came, it was compelled, like any underground resistance, to yield the leadership to regular armies with heavier and more modern equipment."

It is hard to say how much English Protestantism owed to Continental doctrines, and how much to its indigenous Lollard prelude. In the early days, the authorities tended to refer to Protestants more or less interchangeably as "Lutherans" and "Lollards." Bishop Tunstall, in a letter to Thomas More in 1528, warned against the men bringing deviant teaching from Germany and the Low Countries: "There have been found certain children of iniquity who are endeavoring to bring into our land the old and accursed Wycliffite heresy, and along with it the Lutheran heresy, foster-daughter of Wycliffe's." The parts of the country where Lollardy had been strongest—broadly speaking, the southeast of England—were to become the heartland of Puritanism. It was from these rich counties that the greatest numbers of New England's settlers were drawn.

Protestantism was to become critical to the identity of the Anglosphere peoples. It was the main binding agent when England, Wales, and Scotland formed a united nation. The fact that most of Ireland remained Catholic largely explains why the United Kingdom now ends where it does.

Protestantism also bound the peoples of Great Britain to their kindred across the oceans. As late as 1773, Benjamin Franklin was pleading with his fellow colonists not to sunder their links with Britain lest they thereby tip the European balance in the pope's favor: "Remember withal that this Protestant country (our mother, though lately an unkind one) is worth preserving, and that her weight in the scales of Europe and her safety in a degree may depend on our union with her."

This is not a subject about which I write comfortably. In much of the Anglosphere, the association of Protestantism with national identity led to bigotry, violence, and the legal exclusion of Roman Catholics from full civil rights. The two most common derogatory terms applied to Catholics between the sixteenth and nineteenth centuries were *outlandish* and *papist*, intended to stress, respectively, their supposed foreignness and their divided loyalties. Even now, the last echoes of these words yet linger in corners of the English-speaking world: Belfast, Glasgow, Liverpool, even Toronto. Being of Ulster Catholic extraction on one side and Scots Presbyterian on the other, I am perhaps more alert to sectarianism than some of my countrymen, and I have always loathed it. But it is impossible to write meaningfully about the Anglosphere without understanding the worldview of its inhabitants during the early modern period. As the historian Charles Ritcheson put it, "Aside from nationality itself, the Christian religion, and more especially the Protestant variation of it, constituted the greatest common denominator of life in both Britain and America."

In Great Britain, Northern Ireland, North America, and, later, Australia, South Africa, and New Zealand, Protestantism was understood primarily in political rather than theological terms, as a guarantor of free speech, free conscience, and free parliaments. These were not the prejudices of a Whig elite, but deep and popular convictions, regularly refreshed by news of persecution in Europe, nourished by fear of the Spanish Inquisition, and invigorated by the stories of French Huguenots, Flemish Protestants, and other refugees who settled throughout the Anglosphere.

The most widely owned book in Britain in the seventeenth and eighteenth centuries after the Bible was *Foxe's Book of Martyrs*, a grisly account of the tortures and executions visited upon Protestants, especially during the six-year reign of Queen Mary I. Even the poorest classes would buy almanacs, which, alongside the dates for sowing and reaping, listed the anniversaries judged most important in the national story: the break with Rome in 1534; the defeat of the Spanish Armada in 1588; the rebellion in Ireland in 1641; Protestant William of Orange's victory over Catholic James II in 1688; the accession of the Hanoverian dynasty in 1714. One date was doubly sacred: November 5 marked not only the defeat of the Gunpowder Plot in 1605 but also the arrival of William of Orange eighty-four years later.

Anti-Catholic prejudice was not doctrinal—it had little to do with whether one believed in transubstantiation or in praying for the souls of the dead. Rather, it was based, like so many prejudices, on a sense of being under threat.

In 1570, Pope Pius V had called for a rising against the Protestant queen Elizabeth I, and none of his successors had repealed the bull, *Regnans in Excelsis*, that formally deposed her and absolved her subjects of any allegiance to their monarch.

That memory, and that threat, were constant influences upon England's—later Great Britain's—foreign policy. For the next two and a half centuries, the country was in a state of semipermanent war with the Catholic powers of the age: first Spain, then France, occasionally both at the same time. If the popular anxieties of the era had been represented in the form of a map, it would have shown a large arrow stretching from the Continent to Ireland, and two smaller ones jabbing from Ireland to the parts of Great Britain with the largest concentrations of Catholic inhabitants: the Scottish Highlands and Lancashire.

It is easy, with hindsight, to say that these fears were baseless. We know that Britain was ultimately triumphant—emphatically so—in its struggles with its neighbors. But this fact was not apparent to contemporaries, who felt that they were engaged in a life-and-death struggle. As the Earl of Essex told the Privy Council in 1679: "The apprehension of popery makes me imagine that I see my children frying in Smithfield."

The magisterial historian of the seventeenth century, J. P. Kenyon, likened the atmosphere to that of the Cold War, which was still at its height when he was writing. Just as Western communists, even the most patriotic among them, were bound to be seen as potential agents of a foreign power, and just as some suspicion fell even upon mainstream democratic socialists, so seventeenth-century English-speaking Catholics were feared as fifth columnists, and even those High Church Anglicans whose rites and practices appeared too "Romish" were regarded as untrustworthy.

In retrospect, we can see that such suspicions were almost certainly unfounded. Even the most biased Whig-Protestant historians
readily admitted as much. Lord Macaulay, the greatest of them all,
was in no doubt that, had the French or the Spanish invaded, the
Catholic squire would have responded every bit as patriotically as
his Protestant neighbors, holstering his old pistols and marching to
his king's assistance. But that squire's Protestant neighbors, lacking
the benefit of hindsight, were less willing to take the risk.

Were he writing today, Kenyon might have drawn a parallel, not
with the Cold War, but with the status of Islam in the West. Like
English-speaking Catholics in the early modern period, Muslims
are often on the receiving end of a prejudice that is more political
than religious in inspiration. Non-Muslims don't complain about
the practice of hajj any more than Protestants used to complain
about the practice of confession. Most anti-Muslim animosity has
the same root as the older anti-Catholic animosity, namely the fear
that devotees might, in the last analysis, be disloyal to their country.
As John Locke, who believed in toleration for every Christian denomination except Roman Catholicism, put it, "all those who enter
into it do thereby *ipso facto* deliver themselves up to the protection
and service of another prince."

British Catholics eventually overcame these prejudices by making
great play of their patriotism, praying ostentatiously for the monarch
of the day and flying the flag from their churches. By the nineteenth
century, the charge of divided loyalties had been comprehensively
refuted by the long lists of Catholics among Britain's war dead. In
the popular phrase of the time, Catholics had "proved their loyalty."
A similar process will probably take place among Muslims in the
Anglosphere, who will eventually understand that even the most
baseless accusations need to be answered patiently and courteously.

In both cases, those looking to sustain their prejudices could find
the evidence they wanted if they looked hard enough. There really

were foreign wars, and there really were some domestic sympathizers with the enemy, though far fewer than was popularly supposed. As in every age and nation, there were people from the religious minorities who made inflammatory and provocative statements. In a few cases, there were actual conspiracies.

The 9/11 of the seventeenth century was the Gunpowder Plot, which took place shortly after the accession of James VI of Scotland as James I of England. A band of Catholic hotheads, to the horror of most of their coreligionists, planned to blow up the king and his MPs during a parliamentary session, and then launch a general insurrection. Guy Fawkes, the plotter caught with the explosives, was, like Mohammad Atta four centuries later, a religious extremist, a jihadi who had been radicalized in overseas wars. To this day, Fawkes's name is recalled in every corner of the country.

The foreign anthropologist, looking for the chief folk customs of the English, finds few public spectacles as arresting as that of Guy Fawkes Night. Every year, the plotter, very occasionally with the pope alongside him, is burned in effigy on the anniversary of his capture, one of the few dates that English people haven't forgotten:

Remember, remember, the fifth of November
Gunpowder, treason and plot;
We know no reason why gunpowder treason
Should ever be forgot.

Guy Fawkes, Guy Fawkes, 'twas his intent
To blow up the King and Parliament.
Three score barrels of powder below,
Poor old England to overthrow.

In almost every English village, fireworks and bonfires mark the date. Visitors from overseas are often puzzled, and sometimes disgusted, by the spectacle. The English are not an especially spiritual people, and the minority who are regular churchgoers tend to go out of their way to stress their tolerance and ecumenism. Yet here is what looks like an orgy of popular anti-Catholicism of a kind rarely found elsewhere.

The explanation is that few if any of the participants are conscious of the sectarian overtones. Nowadays, Guy Fawkes Night is simply an annual fireworks display, one of the rare occasions when a naturally reserved people will chat happily to strangers. To the extent that people are aware of any political implications at all, they believe they are celebrating the survival of parliamentary democracy, which was secured when the Gunpowder Plot was uncovered. The denominational aspect has been wholly forgotten, and English Catholics now join the festivities in exactly the same spirit as their neighbors.

The anniversary was enthusiastically marked across the Thirteen Colonies before independence, especially in Boston. Then, in 1775, hoping to secure Canada for the revolution, George Washington ordered the suppression of the tradition, which never revived:

> As the Commander in Chief has been apprized of a design form'd for the observance of that ridiculous and childish custom of burning the Effigy of the pope—He cannot help expressing his surprise that there should be Officers and Soldiers in this army so void of common sense, as not to see the impropriety of such a step at this Juncture; at a Time when we are solliciting, and have really obtain'd, the friendship and alliance of the people of Canada, whom we ought to consider as Brethren embarked in the same Cause.

Guy Fawkes Night stands as a neat symbol of what has happened across the Anglosphere, with the exception of parts of the United States: religious observance has declined, but a Protestant *political* culture has not. As the churches have lost their centrality in people's lives, that culture has become simply a set of prevalent Anglosphere values, shared equally by English-speaking Hindus, Jews, atheists, and so on. The association that was almost universally made between religious toleration and political pluralism, between Protestantism and parliamentary democracy, between religious and civic freedoms, is no longer made, but the values it created survive. The ghost has departed, but the machine hums on.

It cannot be stressed too strongly that the Anglosphere's Protestant identity was, at a popular level, predominantly tribal and political rather than dogmatic. Daniel Defoe, the author of *Robinson Crusoe*, talked of "a hundred thousand country fellows prepared to fight to the death against Popery, without knowing whether it be a man or a horse."

Only in Northern Ireland do we still find a widespread sense of Protestant identity that has become separated from church attendance. What is now regarded as a peculiarity of that province is, in reality, the vestigial survivor of a culture that was once common across the core Anglosphere.

It would be a mistake, though, to think that the absence of religious devotion meant that there was nothing left except chauvinism. Chauvinism there certainly was. But there was also a belief—an understandable belief in the context of its time—that the religion of the English-speaking peoples was a guarantor of political liberty. As Sir Henry Capel told the House of Commons in 1679, "From popery came the notion of a standing army and arbitrary power. Formerly the Crown of Spain, and now France, supports the root of this popery amongst us; but lay popery flat and there's an end of arbitrary government and power. It is a mere chimera without popery."

Contemporaries contrasted the parliamentary and constitutional governments of Britain, Switzerland, and the Dutch Republic with the autocratic monarchies of France and Spain. They saw a religious connection and, while their analysis was naturally colored by their own faith, the link was not baseless. The idea that everyone should read the scriptures had egalitarian and democratic implications. Those who strove to, as they saw it, perfect the Reformation by abolishing bishops and allowing congregations to elect their leaders, were consciously campaigning for representative rather than hierarchical government. Their religious convictions were bound to spill over into their political opinions. These groups—Puritans in England, Presbyterians in Scotland and Northern Ireland, Nonconformists and Methodists in Wales, and the coreligionists of all these groups in the New World—provided the core of Whig support down the ages, though that party was sometimes called by different names.

These were, by and large, the communities that championed Parliament against the King in the seventeenth century, and that campaigned for an extension of the franchise in the nineteenth. They were the communities that peopled North America, as English Puritans in New England and as Scots-Irish Presbyterians in Virginia and Pennsylvania. They were the strongest supporters of the American Revolution.

If Britain defined itself as a providential nation, as the preponderant Protestant power in the world and therefore the chief defender of the cause of liberty, early America took that self-definition much further. The Pilgrim Fathers brought with them not only a fear of Catholicism and a distaste for High Church Anglicanism but also a powerful sense of destiny.

The Anglophone expansion into the New World was, from the very beginning, seen in providential terms. As the historian Kenneth R. Andrews put it:

Militant puritanism fused with aggressive nationalism in that fanatical, psalm-singing, image-breaking "cause" espoused by Francis Drake and not a few of his piratical companions. It was in this form, as an ingredient of national feeling, that religion made its chief contribution to the movement of overseas expansion in the period.

The first great propagandist for settling the Americas was Richard Hakluyt, whose *Discourse of Western Planting*, written in 1584, argued that populating the vast and fertile land across the Atlantic would enhance England's wealth, produce useful work for those "lustie youths that be turned to no profitable use," and, above all, secure more souls for Protestantism. Prophetically, Hakluyt foresaw that North America might be a home, not just for English-speakers, but for refugees "from all partes of the world" who were "forced to flee for the truthe of gods worde."

These refugees, when they crossed the seas, were in no doubt that they were making a bargain with the Almighty. In return for deliverance from persecution, they must build a society that allowed no place for what they saw as the superstition, idolatry, and worldliness that had corrupted European churches. John Winthrop, the most famous of all the early Puritan settlers in New England, is remembered today for the sermon in which he told his brethren, "We must consider that we shall be like a City upon a Hill; the eyes of all people are on us." Those words have been quoted by Americans ever since, most famously by John F. Kennedy in his 1961 inaugural address. Yet the passage in Winthrop's sermon that immediately precedes that image gives us a clearer picture of the pilgrim leader's motives:

> We are entered into covenant with Him for this work. We have taken out a commission. The Lord hath given us leave to draw our own articles. We have professed to enterprise these and those accounts, upon these and those ends. We have hereupon besought Him of favor and blessing. Now if the Lord shall please to hear us,

and bring us in peace to the place we desire, then hath He ratified
this covenant and sealed our commission, and will expect a strict
performance of the articles contained in it.

Covenants were enormously important to the Protestant world-
view, informing the development of law and politics throughout the
Anglosphere. Scots Presbyterians, pledging themselves to oppose
the "Romish" practices favored by the Stuarts in the seventeenth
century, called themselves Covenanters. Their alliance with English
Puritans in 1643 was sealed in a document known as the Solemn
League and Covenant—a title copied by Boston radicals in 1774.
In 1912, half a million Protestants in Northern Ireland signed the
Ulster Covenant, swearing to oppose devolution to Dublin. And in
1955, in a little-chronicled assertion of Anglosphere values, thirty
thousand English-speaking South Africans signed the Natal Covenant, rejecting the apartheid state that the Afrikaners were establishing and pledging their loyalty to the Queen and Commonwealth.

The early Americans, however, took the notion furthest, in effect
binding themselves and their descendants to a contract with God.
Implicit in that bargain was a rejection of Catholicism, and, by extension, those Anglican or Episcopalian practices and doctrines that
seemed too close to Catholicism.

American historians were, for a long time, understandably reluctant to stress this element of their national story. The U.S. Constitution enshrined religious toleration, which was an extraordinary idea
in its time. Naturally enough, later Americans preferred to emphasize that side of the story rather than dwelling on the sectarianism
of the Massachusetts congregations who had led the opposition to
George III.

Because our own generation is interested in constitutional issues,
we tend to play up the constitutional aspects of the American Revolution. We tell ourselves that it was all about "No taxation without

representation." And so, indeed, it was. But that is only one part of the story. For many North Americans, the revolution was also a reaffirmation of the religious values that had been carried across the Atlantic by the first settlers. It was the perception of a threat to that heritage that did most to radicalize opinion in the colonies prior to the revolution. The rows over the role of the Church of England in the colonies during the 1760s had soured the atmosphere, making many Americans far readier to quarrel over questions of taxation and trade.

Thomas Secker, who had become archbishop of Canterbury in 1758, was born a Dissenter and, with the misplaced zeal of a convert, was determined to bring the colonies into the Episcopalian fold. Among other things, he backed the establishment of an Anglican missionary church in, of all places, Cambridge, Massachusetts, capital of New England Congregationalism. He tried to get the Privy Council to strike down the Massachusetts Act, which allowed for Puritan missionary work among the Indians. And, most offensive to colonial opinion, he sought to create American bishops.

This last scheme provoked such a fierce backlash that it was promptly dropped by the London authorities. They judged, almost certainly correctly, that the identity of the colonists as Nonconformists or Puritans mattered more to them than their identity as unrepresented electors.

The memory of the original migration, the Puritan hegira, is one of the central grievances recalled in the Declaration of Independence: "We have reminded them [our British brethren] of the circumstances of our emigration and settlement here." Every American understood what those words meant. Theirs was a Protestant nation, founded by men and women who had come to escape the rituals and hierarchy of an only half-reformed English church.

It is hard for the historian to measure which issues most exercised the nonpamphleteering classes. But we can get a sense by looking at petitions, newspaper circulation, and the like. And one action which was almost universally resented in the Thirteen Colonies was the Quebec Act of 1774, by which Britain recognized the traditional rights of the Catholic Church in Canada. For many of the colonists, it was as though the king had sent the serpent into Eden after them. They had come to the New World to get away from Catholicism; now here was their mother government re-creating, as they saw it, popery in their paradise. The Quebec Act, too, is referred to angrily in the Declaration of Independence: "For abolishing the free System of English Laws in a neighboring Province, establishing therein an Arbitrary government, and enlarging its Boundaries so as to render it at once an example and a fit instrument for introducing the same absolute rule into these Colonies."

George III, for his part, was in no doubt as to what the conflict had been about. To the end of his days, he referred mournfully to the loss of the colonies as "my Presbyterian war."

Every nation, of course, cherishes its foundation story. And there is nothing mythical about the recognition of religious freedom, both as a principle in the Bill of Rights, and as an enduring characteristic of the American Republic ever since. Yet we cannot avoid the evidence. The American Revolution was, at least in part, the result of a spasm of religious intolerance. That this spasm should have engendered the first truly secular state on earth, one in which all religions might compete on even terms, is close to miraculous.

THE SAME IDEALS

Every American schoolchild knows the story of Paul Revere's ride. The Boston silversmith was one of those plucky Patriots who, in April 1775, dashed through the night to warn their neighbors that General Thomas Gage and his soldiers had slipped secretly from their barracks and were on their way to disarm local militia. The names of the other riders are now forgotten. Revere, a prominent local businessman and Freemason recognized everywhere he went, was the one whose warnings were heeded.

Through the small communities of eastern Massachusetts he thundered, in a gallop now reenacted every year: Somerville, Medford, Arlington, Lexington, Concord. Wherever he went, as any American will tell you, he roused local Patriots to arms with his warning: "The British are coming!"

Except that, in reality, he shouted no such thing. It would have been extremely eccentric to yell "The British are coming!" at a population that had never thought of itself as anything other than British.

What Paul Revere in fact shouted was "The regulars are out!" (Or, according to one source, "The redcoats are out!") In America, as elsewhere in the Anglosphere, people had an ingrained distrust of standing armies, seeing them as instruments of internal repression.

This was the very issue that had triggered the English Civil War in 1642—a conflict constantly in the minds of New England Puritans, the ideological and in many cases also the familial heirs to England's Roundheads.

That war had been about many other things, too, of course. But it was Charles I's attempt to command the militia that had triggered the fighting. His opponents had been determined to wrest control of land forces away from the Crown, and they eventually succeeded—which is why, to this day, the United Kingdom has a Royal Navy and a Royal Air Force but a *British* Army.

History was about to repeat itself. The day after Paul Revere's ride, April 19, 1775, the first shots were fired in what most contemporaries took to be the second English Civil War—a continuation, in many ways, of the first. No one at that stage thought of the conflict as being between two distinct nations.

To describe the fighting at Lexington and Concord as being a battle between British and American forces, as the tour guides do, is plain wrong. It was a battle between supporters and opponents of George III's ministry—a division that cut laterally across American and British opinion.

To assume the preexistence of an American political consciousness—to imagine that either Revere or anyone else could have spoken in 1775 of "the British" as a foreign people—depends not only on ignoring how contemporary Americans spoke and wrote but on disregarding the text of the Declaration of Independence. We have already noted the way that document decries the abolition of "the free System of English Laws" in Quebec. But the signatories had a more immediate grievance to lay against George III: "He is at this time transporting large Armies of foreign Mercenaries to compleat the works of death, desolation, and tyranny." Foreign mercenaries: in other words, soldiers who were not British. The hapless king's decision to turn Hessian and other German hirelings against the British population of America was final proof, in the eyes of the revolutionaries, that they had lost their status as Britons.

Of course, once the revolution was victorious, its beneficiaries did what all successful revolutionaries do: they backdated their cause, reinterpreting the recent past as if there had been an American nation all along—a nation in arms against a colonial power.

It was in this spirit that the poet Longfellow immortalized Revere in schmaltzy verse:

> *Listen my children and you shall hear*
> *Of the midnight ride of Paul Revere,*
> *On the eighteenth of April, in Seventy-five;*
> *Hardly a man is now alive*
> *Who remembers that famous day and year. . . .*

By the time those words were written, nearly a century after the events they purported to describe, the United States had been an independent power for so long that it was hard to think of the revolution as having been an Anglosphere civil war, a war that had pitted Whig against Tory rather than American against Briton. Yet, as we shall see, public opinion in Great Britain, at least until the French became involved, seems to have been almost identical to that in the colonies, with perhaps 30 or 35 percent of the population broadly Tory in sympathy. The difference was that the colonial assemblies were elected on a wider franchise, and so were more representative of public opinion than the unreformed House of Commons.

We shall look in more detail at the American Revolution later on. For now, it is enough to note that, when Winston Churchill talked about the two nations having "more or less the same ideals," he wasn't making some bland generalization about being the good guys. He was keenly aware that the principles Britain was defending in its struggle against the Nazis were the same as those on which the American Republic had been founded.

As Churchill was to put it in his *History of the English-Speaking Peoples*, "The Declaration [of Independence] was in the main a restatement of the principles which had animated the Whig struggle against the later Stuarts and the English Revolution of 1688." Indeed it was, often in the most literal way: the right of petition, the prohibition of standing armies, the protection of common law and jury trials, the right to bear arms—all were copied from England's revolutionary settlement of 1689. Some of the clauses of England's 1689 Bill of Rights were reproduced without amendment. Here, for example, is the English Bill of Rights on criminal justice: "Excessive bail ought not to be required, nor excessive fines imposed, nor cruel and unusual punishments inflicted." And here is the U.S. Constitution: "Excessive bail shall not be required, nor excessive fines imposed, nor cruel and unusual punishments inflicted." Both these clauses themselves looked back to Magna Carta. Both were understood by their authors not as the creation of a new privilege, but as the confirmation of an ancient one.

Indeed, the more we see of the commonalities between American and British constitutional documents, the more meaningless it is to talk of the one having copied the other. Both are expressions of an inherited folkright of constitutional freedoms—a folkright that was and is the common property of Anglosphere societies.

If Paul Revere's imagined cry was an invention of later writers, what are we to make of the word *patriot*, in use in the American colonies long before anyone dreamed of a rupture with London?

During the second half of the eighteenth century, the word had the same connotations in America as in Great Britain. A Patriot was someone with a pronounced devotion to liberty and property; someone who stood for the interests of the nation as a whole, rather than those of its supposedly effete ruling class. Yet, before 1776, there was no American nation. An American's loyalty to his colony was contained within a wider sense of allegiance to the British imperium. To what, then, was he referring when he called himself a Patriot?

The answer is that he saw himself as a *British* patriot, standing up for his inherited freedoms against those seemingly bent on eroding them—namely an autocratic King and his fawning ministers. By patriotism, the American meant being loyal to the people around him. He meant sticking up for his community instead of fawning to the authorities. He meant being prepared to forgo the salaries, pensions, and sinecures that might have come his way had he been readier to ingratiate himself with his governor and the colonial administration.

To put it more cynically, "Patriot" was the name that Whigs, on both sides of the Atlantic, gave themselves. In doing so, they were consciously stressing the unique political inheritance of the

Anglo-American empire—common law, Magna Carta, the English Bill of Rights—and implicitly accusing their opponents of favoring illiberal, authoritarian, and foreign alternatives. Patriots loudly asserted—and not without justice—that personal freedom and resistance to autocratic power were the distinguishing features of the English-speaking peoples. In the words of "Rule, Britannia," composed in 1742:

> *Though nations not so blessed as thee*
> *Must in their turn to tyrants fall*
> *Thou shalt flourish great and free*
> *The dread and envy of them all.*

All political factions, of course, contain their share of careerists and adventurers. There were some who called themselves Patriots yet were prepared to put personal gain before the national interest. These were the men Dr. Johnson had in mind when he made his best-known, and most misunderstood, remark: "Patriotism, Sir, is the last refuge of a scoundrel."

Johnson's biographer James Boswell, who recorded the quip, immediately added "but let it be considered, that he did not mean a real and generous love of our country, but that pretended Patriotism which so many, in all ages and countries, have made a cloak of self-interest." Johnson, a partisan Tory, was mocking the men he once called "Whig dogs." For all their talk of liberty, he was saying, they were as venal as anyone. He was, in other words, using *patriot* in its contemporary sense, as the name of a political clique—albeit doing so pejoratively.

American Loyalists used the word in much the same way. "Patriots," in the eyes of that large minority in the colonies who favored the constitutional status quo, were opportunists, using the language of liberty as a cover for their own ambitions. Some, they alleged, were trying to escape their London creditors. Others were smugglers. Still others were unscrupulously whipping up the mob with talk of equality when their real aim was to replace a Tory colonial elite with a Whig one.

Neither side used *patriot* to mean someone who favored American over British interests. Not until much later did novelists and scriptwriters start pretending that the word had been employed that way. The notion of an exclusively American patriotism was born after 1776. Before then, a patriot was someone, on either side of the Atlantic, who was determined to preserve the libertarian exceptionalism of the English-speaking polity against its enemies, internal or external.

The Virginia-born Lady Astor described the American War of Independence as having been fought "by British Americans against a German King for British ideals." The description of George III as German was a little unfair—unlike the first two Georges, he had been born in Britain, spoke English as his first language, and, as he liked to put it, "gloried in the name of Briton." Perhaps, in the context of the time, Lady Astor could be forgiven the xenophobic barb: she was speaking as Luftwaffe bombs were raining on London; and she had, in any case, become rather eccentric at that stage in her life. Still, she was onto something. The Patriots of the 1770s were indeed fighting for British ideals, just as their descendants were to do in Lady Astor's time.

These ideals—free elections, habeas corpus, an uncensored press, and the rest—are now universally acknowledged, if by no means universally applied. It is easy, these days, to be complacent, to assume that every nation moves in the direction of liberal democracy when it becomes wealthy enough, educated enough, and secure enough.

It takes a real mental wrench to see the world from the perspective of August 1941, when the vast bulk of humanity lived under totalitarian governments. Almost the entire Eurasian landmass, from

Brest and Lisbon to Seoul and Vladivostok, was under one form of dictatorship or another—fascism, communism, or Japanese militarism. Freedom, the rule of law, and democracy were, to a single approximation, confined to the Anglosphere. When Churchill spoke of Britain and America having the same ideals, he knew that those ideals existed almost nowhere else. Their continued existence depended on the military victory of the English-speaking peoples.

Nor, in 1941, could you have consoled yourself with the thought that constitutional government had simply been temporarily displaced, trodden under the boots of invading soldiers. In a tiny number of nations—the Low Countries, France, and Scandinavia—the Nazis had overturned parliamentary regimes. But these regimes had, by the 1930s, become the exception. The list of countries that had turned to authoritarianism domestically, without needing to be invaded, was far longer: Austria, Bulgaria, Estonia, Greece, Hungary, Italy, Latvia, Lithuania, Poland, Portugal, Spain, Romania. Much of Latin America, too, was under one form of dictatorship or another.

No one in the 1930s saw democracy as a coming force. On the contrary, it had been ousted from state after state. Fascists and communists used remarkably similar language when they talked of parliamentary regimes. The old democratic-capitalist order was almost always described as "rotten" or "decadent."

"Capitalism has run its course," Hitler told an enthusiastic Mussolini in 1934, using precisely the same language as Marxist-Leninists. Both groups saw liberalism as a perversion of the natural order, in which the group was supposed to be more important than the individual. Personal freedom within a parliamentary regime was seen as a temporary aberration, a hiatus during which the "money-grabbing classes" had established an artificial hegemony.

Consciously distancing themselves from what they saw as the old order, authoritarians of left and right used the imagery of renewal,

youth, revolution, a fresh start. Fascists and communists presented
themselves as the only viable alternative to each other. Bourgeois
democracy, they agreed, was finished. Tomorrow belonged to them.

We now know, of course, that what the Nazis called "decadent
Anglo-Saxon liberalism" was not in decline at all. On the contrary,
it would emerge from the ideological wars of the twentieth century
as the most successful system on earth. But in 1941, free-market democracy still looked like what it was: a political ideology that was
confined to the Anglosphere. Most observers assumed that the statist ideologies, which elevated martial vigor, collective endeavor, and
self-sacrifice, were bound to triumph over the comfortable, bourgeois values of the Anglo-Saxon liberals.

Reinhard Spitzy, private secretary to the German foreign minister, Joachim von Ribbentrop, later lamented that that assumption
had been wrong: "We Nazis never said we were nice democrats. The
problem is that the British seem like sheep or bishops, but when the
moment comes they are shown to be hypocrites, and they become
a terribly tough people." Wherein lay the hidden strength of the
English-speaking democracies? How did a group of people derided
by their enemies as decadent, materialistic, lost to all considerations
of faith or honor, manage to triumph over all their enemies? What
was the Anglosphere's secret?

CHAPTER 2

Anglo-Saxon Liberties

Let me state a certainty. Late Anglo-Saxon England was a nation-state. It was an entity with effective central authority, uniformly organized institutions, a national language, a national church, defined frontiers (admittedly with considerable fluidity in the north), and, above all, a strong sense of national identity.

—JAMES CAMPBELL, 2000

The English law existed not to control the individual but to free him.

—ROGER SCRUTON, 2006

WHO ARE THE ENGLISH?

We shall never know who the first Englishmen were, or why they came. Perhaps they landed as traders, exchanging the rude produce of their forest homes for the advanced manufactures of Roman Britain. Perhaps they arrived as soldiers of fortune, drawn by the prospect of steady pay and the plot of land traditionally given to legionaries after twenty years' service. Possibly, even in those very early days, they came as marauders: pillage had always been an attractive alternative to hacking a living from cold northern soil.

Whatever their motives, they brought with them something more precious than any booty a Germanic war chief could conceive. They had a way of ordering their affairs that was already diverging from the customs of other peoples. They were developing a notion of the relationship between a man and his tribe that was to become, in time, the greatest export of English-speaking civilization, its supreme contribution to the happiness of mankind. In the damp green island that was their new home, they would evolve theories of kingship and government, of property and contract, of law and taxation, that were to transform and elevate our species.

At the time, such a notion would have seemed preposterous. Britannia was among the more orderly and prosperous late Roman provinces. The Teutonic tribes who raided from across the North Sea were viewed with that peculiar mixture of scorn and fear that higher civilizations direct at savages.

But the power of those tribes was waxing as Rome's waned. At first, the military authorities built a chain of forts along Britain's eastern coastline to protect the fertile flatlands from seaborne barbarians. But the raids became more frequent as the resolve of the defenders wavered. All over Europe, Germanic tribes, whether as rebelling mercenaries or as invaders, were overturning the Roman order. In AD 410, the legions were recalled to defend the imperial capital, and the province of Britannia was left to fend for itself.

The largely Christian population found itself assailed from three directions: Irish clans attacked from the west, Angles and Saxons from the east, Picts from the north. It may well be that the Romano-Britons paid one of their foes to defend them from the other two, hiring bands of Anglo-Saxon mercenaries. That is the traditional account, though the earliest sources we have were written much later.

What is certain is that, during the fifth century, the fathers of the

English—that is, tribes from what are now Germany, Denmark, and the Low Countries—began to settle in large numbers. By now there was no question about their motives. They came for territory. The numbers are almost impossible to establish. Anything between 20,000 and 200,000 people might have left their tangled forests for the new land during the fifth and sixth centuries, bringing with them the same grave-artifacts that are found along the Elbe. And bringing, too, the germ of what later generations would call Anglo-Saxon liberties.

What happened to the existing population of what is now England? Was it exiled or absorbed? The question matters, not least because it touches on whether Anglosphere values were, right from the beginning, developed in a multiethnic context. The debate still divides historians, though the consensus has shifted markedly over the past decade.

It was once believed that the Anglo-Saxons almost entirely displaced the natives. Surviving Britons were said to have been driven into Wales, Cornwall, northwest England, and southwest Scotland—as well as Brittany, which they conquered in their turn.

Two main pieces of evidence were advanced in support of this interpretation. The first was linguistic. The tongue of the Britons lived on in the western part of the island, which the Anglo-Saxons were late in reaching. One version, known as Cumbric, was heard until the eleventh or twelfth century in the British kingdom of Strathclyde, mainly in what is now southwest Scotland. Another, Cornish, didn't entirely die out until the nineteenth century. Welsh is still spoken by around a fifth of the population of Wales.

Yet the common tongue that fathered these related idioms (the Brythonic languages, as linguists call them) left almost no trace in English. Had there been a racial intermingling, some historians maintain, we would expect to find a number of Celtic-derived words in our speech. In fact, there are almost none. And those few that there are—*crag, tor, combe*—tend to be words for geographical features found only in the northern and western parts of Great Britain: the parts that were absorbed last into the English-speaking kingdoms, and where the Anglo-Saxon influence was weakest.

The second piece of evidence is the sole work of British history surviving from the sixth century: *De Excidio et Conquestu Britanniae*, "On the Ruin and Conquest of Britain," written by a Welsh monk called Gildas.

Gildas describes a genocide. He quotes a letter that the Britons sent pleadingly to Rome, describing their plight: "The barbarians drive us to the sea, the sea pushes us back at the barbarians: between these two deaths either we are drowned or our throats are cut." Gildas, however, was writing a hundred years after the events he described. He was, moreover, writing with an agenda. Like most chroniclers of the period, he believed that God rewarded and punished whole nations. As Jehovah had unleashed the Assyrians against the Israelites, so He was now raising up pagan invaders to chastise Britain's Christians for falling into sin. It suited Gildas's purpose to treat the Anglo-Saxons as a divine scourge, a pitiless and unstoppable force.

Common sense suggests a more nuanced picture. Not only would some Britons have remained on their land as conquered vassals; some, almost certainly, would have allied with the newcomers. We know that leagues between Welsh and English chiefs were common in later centuries, and similar things happened across Europe where Germanic war leaders made pacts with local magnates. Population displacement typically occurs when there is a significant technological disparity between two peoples, which, in this case, there was not.

As for the linguistic evidence, there is no reason to suppose that the absence of Brythonic vocabulary in English signifies the extirpation of its speakers. Consider, as a modern analogy, my native Peru. Around 40 percent of the population there is of indigenous origin,

40 percent mixed, and 20 percent largely or wholly white. A portion of the native peoples—some 15 percent of the total population—still speak their aboriginal languages, Quechua and Aymara. The rest, including almost the entire mestizo population, speak Spanish. Peruvian Spanish is unadulterated—purer, in many ways, than the Spanish of contemporary Iberia. It contains as few Quechua-derived words as English contains Brythonic ones.

A future historian, working with nothing more than the linguistic evidence, might well conclude that the indigenous Peruvians had been slaughtered or driven out. He would be wrong. Although they adopted the customs and mores, the language and law, the religion and identity of the conquerors, they remained, and remain, the racial majority. It is not hard to imagine something similar having taken place in fifth- and sixth-century England.

Even before it became possible to sample DNA, many historians disputed the annihilation thesis. Place-name evidence suggested some British survival even in the Germanic heartland. For example, several English towns have the stem-name *eccles* (Ecclesbourne, Eccleston, and so on), which derives from the Latin word for church. There is an Eccles-on-Sea in Norfolk, raising the possibility that, even at the easternmost edge of England, there was a Christian (and therefore Romano-British) community large enough to sustain a place of worship.

Some English place-names are tautological, using Brythonic and Anglo-Saxon words for the same thing. In Buckinghamshire, for example, there are two nearby villages called Brill and Chetwode. Brill takes its name from the Celtic word *breg*, meaning hill: Breghyll became Brill. Chetwode is derived from the Celtic word *cet*, meaning woodland. The bastard place-names thus literally mean "Hillhill" and "Woodwood."

It's possible to infer from these names that the invading Saxons took such little interest in the aboriginal people that they didn't even pick up their basic words for "hill" or "wood," and so assumed that Breg and Cet were names for a particular hill, a particular wood. But it's surely also plausible to see them as evidence of the kind of joint naming that happens in places where separate languages coexist. Look at the road signs in, say, Belgium, which refer to towns by their often very different French and Dutch forms: Mons/Bergen, for example. French and Dutch remain quite distinct, with only a few loanwords in the border regions, and these joint names advertise the presence of two language groups.

(Fans of *The Lord of the Rings*, incidentally, might find the names of Brill and Chetwode familiar. When the hobbits leave the Shire, the first place they visit is Bree under Breehill, lying by the Chetwood. J. R. R. Tolkien, a professor of Anglo-Saxon at Oxford, knew a thing or two about English toponymy.)

For a long time, the racial evidence was inconclusive. Even now there is no unanimity. Some historians still adhere to the annihilation narrative, though a growing number incline to a belief in intermingling. Over the past fifteen years, the relatively young field of population genetics has started to give us a clearer picture.

Most studies suggest that the modern English descend, not just from the pre-Saxon population, but from the pre-Celtic. The genetic sequences of contemporary Britons have a strong affinity to those of the Stone Age population. There was much excitement in 1997 when the mitochondrial DNA of "Cheddar Man," the oldest skeleton in Britain, excavated near Cheddar in Somerset, and dating from 7150 BC, was found to match two of Cheddar's present residents.

In 2007, having led a major study with Oxford University that tested more than six thousand samples, genetics professor Bryan Sykes concluded that only 10 percent of men "now living in the south of England are the patrilineal descendants of Saxons or Danes . . . that figure rising to 15 percent north of the Danelaw and 20 percent in East Anglia." Another geneticist, Stephen Oppenheimer, argued that

68 percent of English DNA predates the first farmers in the fourth millennium BC, and that most of it came via Iberia.

Other studies, especially those based wholly on Y chromosome DNA (that is, the genetic material carried only in the male line), have implied a much more significant Anglo-Saxon influx. A major analysis in 2002 found that men in central England were genetically almost indistinguishable from men in Frisia on the Dutch coast, but markedly different from men in Wales. A project by University College London in 2011 reached similar conclusions, finding that half the men in Britain shared a segment of Y-chromosome DNA that is almost universal in Denmark and northern Germany.

Is there a way to reconcile these data? It's a developing science and there are methodological differences in how the studies were conducted. Still, one conclusion that tends to emerge is that Y chromosome DNA surveys (that is, those recording only father-to-son descent) show a heavier Germanic trace, from which finding we can perhaps infer that Anglo-Saxon men took native wives—a pattern common to migration in every age and nation. We can reasonably conclude from the evidence, too, that Germanic ancestry becomes more diluted as you travel from east to west across Great Britain.

As we shall see, later generations of Englishmen were interested in their racial origins—though never, thank God, in the way that some of their Continental contemporaries were. During the English Civil War, many supporters of the parliamentary cause liked to see themselves as authentic Anglo-Saxons struggling to throw off an effete, Norman-descended aristocracy. In their minds, law, liberty, and representative government were part of their birthright as Anglo-Saxons, a patrimony that had come down to them from the tribal parliaments of primeval Germany described by Tacitus.

"Our progenitors that transplanted themselves from Germany hither did not commixe themselves with the ancient inhabitants of the country of the Britons," argued the Roundhead leader John Hare in 1640, "but totally expelling them, they took the sole possession of the land to themselves, thereby preserving their blood, laws and languages uncorrupted."

Similar arguments were made by some of the American Patriots at the time of the revolution. Thomas Jefferson saw Americans as true Anglo-Saxons, who had carried their freedoms into the New World and preserved them there in a purer form than in the old country. (Benjamin Franklin, it must be said, took a very different view, having a low opinion of Germans.)

The idea of a folkright of freedoms, stretching right back to the primitive forest assemblies of the earliest German tribes, animated writers throughout the Anglosphere right up until the 1930s, when the Nazi madness simultaneously discredited racial theory and made unfashionable the idea of kinship with Germans.

From what we can now see of the genetic evidence, such theories were, if not baseless, certainly exaggerated. The English were never a wholly Germanic race. The earliest Anglo-Saxons mingled, at least to some extent, with the ancient Britons. Their descendants were later to mingle with Danes, Normans, Flemings, Huguenots, Jews, Kashmiris, Bengalis, Jamaicans—and, in the New World, with peoples from every continent and archipelago. Intellectual exchange, rather than insemination, turned out to be the surest way of passing on ideas of how to organize society.

The multiethnic nature of the Anglosphere is not a modern phenomenon. In 1703, Daniel Defoe published *The True-Born Englishman*, which pulses with a self-mockery that few contemporary nations felt secure enough to indulge:

Thus from a mixture of all kinds began
That het'rogeneous thing, an Englishman:

In eager rapes and furious lust begot,
Betwixt a painted Briton and a Scot.
Whose gend'ring off-spring quickly learn'd to bow,
And yoke their heifers to the Roman plough:
From whence a mongrel half-bred race there came,
With neither name, nor nation, speech nor fame.
In whose hot veins new mixtures quickly ran,
Infus'd betwixt a Saxon and a Dane.

Anglo-Saxon values, as Richard Dawkins might put it, are a meme rather than a gene. They can be transmitted without any genetic vehicle. They explain why Bermuda is not Haiti, why Singapore is not Indonesia, why Hong Kong is not China (and, for that matter, not Macau).

These precepts had their genesis in the earliest Anglo-Saxon settlements, in the dark years, violent and unchronicled. From that era came three interrelated concepts that were to transform humankind. First, the idea of personal autonomy, including in contract and property rights; second, the notion that collective decisions ought to be made by representatives who are answerable to the community as a whole; third, the conception of the law as something more than a projection of the wishes of the ruler, as a folkright of inherited freedoms that bound the King just as surely as it bound his meanest subject.

Anglo-Saxon values made possible the transformation of our planet over the past three centuries, allowing extraordinary numbers of people to enjoy an unprecedented standard of living. Nowadays, those values are so widespread that we can easily think of them as an inevitable outcome of human development. It takes a major effort of will to imagine quite how revolutionary they must have seemed when first proposed.

FROM WITAN TO WATERGATE

When Bill Clinton narrowed his famously seductive eyes and told television viewers, "I did not have sexual relations with that woman," he was defending himself against a procedure invented by the early English as a way to ensure that even the biggest man in the territory couldn't bend the rules.

Impeachment had been used against some of the most powerful men in the kingdom, from Lord Latimer, impeached for corruption and collusion with the French in 1376, to Lord Melville, the war secretary, acquitted but effectively ruined after being impeached in 1806 for misappropriating public funds.

When congressional Republicans initiated proceedings against President Clinton, they weren't invoking a device that happened to resemble that used in medieval England. It was precisely the same procedure, carried by Englishmen to Massachusetts and Virginia, enshrined in the early colonial charters, and later written into Article I of the U.S. Constitution.

Impeachment is an exceptional recourse, its rarity reflecting its gravity. English (and later British) MPs would often let decades, even centuries, pass without turning to it. The exception was the turbulent period between 1640 and 1642, when, after an eleven-year absence, Parliament was recalled by a desperate Charles I. The MPs of that era had amassed many grievances against their government during what they called the "eleven-year tyranny," and they were determined to seek justice. Fully a quarter of all the impeachment cases heard in England date from those two years, when parliamentary leaders, consciously reviving what they saw as an ancient redress against autocratic rulers, launched proceedings against several of Charles I's ministers.

The nostalgic Anglo-Saxonism of those MPs was a touch Romantic, but not entirely wrong. The idea that an assembly might remove an official—even a king—who had broken the law did indeed have pre-Norman roots. As we shall see, the Anglo-Saxon assemblies, the Witans, had on occasion rejected the claims of their monarchs on grounds of what would now be called abuse of office.

Ponder, for a moment, that astonishing fact. More than a thousand years ago, in England, the precedent had been set that a ruler might be judged before a representative assembly. The law, in other words, was not simply the sovereign's decree; nor yet was it an interpretation of Holy Scripture. The law, rather, was a set of inherited rights that belonged to every freeman in the kingdom. The rules did not emanate from the government, but stood above it, binding the King as tightly as they bound the poorest ceorl. If the monarch didn't uphold the ancient laws and customs of his realm, he could be removed.

English-speaking peoples still commonly, and exceptionally, talk of "the law of the land." Not the King's law, not God's law, but the law *of the land*—a set of rights and obligations immanent in the country, growing incrementally, passed down as part of the patrimony of each new generation.

Of course, the law of the land presupposes a land: a recognized nation to which all freemen belong, and which has an organic identity that amounts to more than simply being a territory under a single government. If the sovereign himself is required to keep that law, it must have a higher source of legitimacy. And that legitimacy resides in the notion of inherited rules belonging to the nation or, in the phrase used by English-speaking peoples when they took to the field against kings who were deemed to have broken these rules— King John, King Charles I, King George III—"immemorial laws and customs."

The story of the development of the rule of law in England, and of all the freedoms that, in consequence, spread across the Anglosphere, is thus the story of the development of England itself as a recognizable unit—as, in other words, a nation.

THE FIRST NATION-STATE

It is extremely rare to find justice, freedom, or representative government flourishing in any context other than a nation-state. All
over the world, we see migration from states with arbitrary frontiers to states that contain more or less homogenous nations. The
rise of the nation-state was a critical element in the success of the
West from the eighteenth century onward. Yet, quite astonishingly,
the process had taken place in England centuries earlier. From England's national unification, and the sense of civic engagement that
followed, came truly breathtaking developments.

Historians tend to think of the formation of the nation-state as
an early modern phenomenon. In most of Europe, it was closely
linked to the cause of democracy.

When, in the eighteenth and nineteenth centuries, radicals
began to agitate for the idea of government of, by, and for the people,
they found that they had immediately raised the question "What
people?" Within what unit, in other words, was the democratic process to be played out?

The answer they offered was, in truth, the only possible one.
Representative government works best when people feel enough in
common with one another to accept government from each other's
hands. Democracy requires a "demos": a unit with which we identify when we use the word *we*. As Charles de Gaulle was to put it,
broadcasting to occupied France from London in 1942, "Democracy
and national sovereignty are the same thing."

Most democrats were nationalists in the sense that they wanted
to align the units of government with the wishes of the people
being governed. Until the nineteenth century, much of Europe was
a patchwork of dynastic territories, created by conquest, marriage,
and happenstance, and corresponding only incidentally with the

preferences of local inhabitants. Democrats wanted to replace these arbitrary states with national units defined by common identity.

Multinational states, they argued, would never be properly democratic. Without national consciousness, people's loyalty would be to something other than the state. As long as a large chunk of the citizenry didn't want to belong at all, there would be repressive measures, and both democracy and liberty would suffer.

It's certainly true that most of the big multinational states in Europe—the Habsburg, Romanov, and Ottoman monarchies, for example—were autocratic. Once their constituent peoples were given the vote, they tended to opt for national separation.

It doesn't follow, of course, that all nation-states are democratic; far from it. It's simply that the alignment of state and national frontiers creates the circumstances where representative government *might* develop.

Historians and political scientists are familiar with these arguments in the context of nineteenth-century Europe. But, perhaps fearful of seeming anachronistic, they never think to apply them to pre-Norman England.

Yet the development of parliament in Anglo-Saxon England— and in a handful of related, homogenous states, notably Denmark and Iceland—anticipated representative government in Europe by several centuries. This remarkable head start owed a great deal to the precocious emergence of a recognizable nation-state in England.

Italians didn't fulfill their national aspirations until 1861, when Garibaldi handed a united peninsula (minus, for the time being, Rome and Venice) to King Victor Emmanuel II. Germans had to wait until 1871, when, in the exquisite Hall of Mirrors at Versailles, a group of princes and generals, resplendent in their stiff uniforms, proclaimed Wilhelm to be emperor of a second German Reich. Farther east, many European nations were under one form or another of foreign rule until 1918.

Yet the English had, on almost any definition, formed an independent and unitary nation-state by the tenth century. No other European country came close. Denmark, arguably the second, coalesced between the eleventh and thirteenth centuries. Iceland, colonized from the eleventh century and also separated by sea from its neighbors, was another early outlier. A case can perhaps be made for Portugal, which showed signs of national consciousness from the twelfth century onward. Most European nation-states, however, emerged in the age of muskets rather than that of battle-axes.

It's important to define what we mean by nation-state. By the tenth century, the people of England had a palpable sense of common identity. As the historian Susan Reynolds put it: "The inhabitants of the Kingdom of England did not habitually call themselves Anglo-Saxons (let alone Saxons), but English, and they called their kingdom England. It was not a hyphenated Kingdom, but one whose inhabitants felt themselves to be one people."

That identity sustained a unitary government whose legitimacy was acknowledged across a defined territory, and whose writ ran unchallenged. Tenth-century England was, by contemporary standards, affluent and powerful, maintaining a uniform coinage and, albeit with some regional variations, a common legal jurisdiction. It had had, since 669, a national church, with every see acknowledging the supremacy of the dioceses of York and, above all, Canterbury.

It's true that, as in any nation-state, things were not entirely clearcut. The Anglo-Saxons were continuing to expand westward, and the conquered Britons were called, by the English, "foreigners" or "Wealas," which later became "Welsh." There was a gradual process of assimilation, during which the Welsh in England were treated as a conquered race, not as full citizens: law codes recognized them as a separate category. On the other side of the country, there was also a gradual assimilation of the descendants of Danish invaders. And the northern border was fluid, with Scottish kings periodically accepting the suzerainty of their southern neighbors, and with an English-speaking population in southern Scotland expanding substantially in the eleventh century.

Every nation-state, however, has marginal ambiguities of this sort. James Campbell's words, which open this chapter, stand. The English were a people: "gens Anglorum," as the historian Bede put it in the early eighth century, "Angelcynn" or "Engelcynn" in their own tongue. And, long before any other Western nation, unless we count pre-Roman Israel, they had turned their nation into a state.

This unique development made possible the other peculiar characteristics of early England. There was not, as in most of Europe, a semi-independent hereditary aristocracy. The law was applied nationally by the King's courts, not locally by territorial magnates (except possibly in the extreme north, where the evidence is patchy). Instead of running semi-autonomous dukedoms, powerful men would meet in a national council, which was in time to develop into a national parliament. Unity, combined with island geography, meant internal order. War bands were mobilized only when there was a war. The distrust of standing armies, which was to be such a powerful feature of later Anglosphere politics, can be traced right back to the organization of the Anglo-Saxon county militia, the fyrd.

England's relative stability gave its inhabitants security in the tenure of their property. Because the state's authority was undisputed, and the regime unlikely to be overthrown, men were commensurately ready to take their disputes to court instead of settling them by force. A unique legal system grew up, based on precedent rather than statute, and with rules on contract and ownership that differentiated England from most of Europe and Asia. It was a legal system that, as we shall see, later engendered modern capitalism.

Nation-states are remarkably secure vessels for freedom. The common identity of their citizens, their shared loyalty to a single authority, their patriotism, tends to enlarge civic society and reduce the need for state coercion.

These aspects of English particularism were to continue and, indeed, to strengthen as England merged with the other territories that make up the British Isles. Between the sixteenth and eighteenth centuries, Britain became a nation-state with (outside the Scottish Highlands and North Wales) a sense of identity as strong as that which had underpinned the rule of law in England. The same unique aspects of English-speaking civilization—common law, independent courts, representative assemblies, sanctity of contract— were carried overseas by Anglophone settlers and administrators. They were maintained in many former colonies after independence. Since so much follows from the early development of a nation-state in England, it is worth looking at how that development occurred.

THE MAKING OF ENGLAND

The early Germanic settlers came in kin groups and tribes, reflecting their varied origins on the Continent. These units eventually coalesced into distinct kingdoms. A thousand years later, British historians would refer to these kingdoms as "the heptarchy"—a phrase that remained current well into the mid-twentieth century, and which, even now, is not wholly archaic. But we shouldn't imagine that there were seven neatly defined Anglo-Saxon realms. Borders shifted constantly. One regime might assimilate another. A conquered kingdom might retain a measure of identity and, after a while, reemerge under its old name.

At the end of the sixth century, there were twelve kingdoms. These later merged into seven, and by the time the Viking invasions began at the end of the eighth century, they had been consolidated into four: Northumbria, Mercia, East Anglia, and Wessex.

Long before the political unification of England, there had been some notion of hegemony over the totality of the English people. Sources refer to occasional overlordship of the English— ambiguously from the early seventh century, and definitively from the early eighth. The chronicles give that position a title: *bretwalda*— almost certainly a scribal mistranscription of *Brytenwalda*, Britain-Ruler. A *bretwalda* was a high king who had managed, for a time, to claim suzerainty over the other English monarchs.

These monarchs didn't regard each other as foreign. They were evidently fond of genealogy, and would promulgate lengthy family trees, tracing their descent back through various early Anglo-Saxon kings to the ancient English gods, Dunor and Wotan. When they embraced Christianity, they cheerfully tacked on an extra few generations, stretching from Wotan through Noah to Adam. The purpose of such bloodlines was, in part, to emphasize their common kinship—a kinship they recognized with each other, but not with neighboring Irish, Welsh, Frankish, or Danish chieftains. Long before anyone thought of England, there existed a sense of an English people—"Anglii" to the scribes who wrote in Latin, "Angelcynn" to those who used the vernacular—defined above all by language. As in Italy and Germany more than a thousand years later, a common language and identity was the prelude to common political institutions.

In Germany, it was Prussia that led the unification process, in Italy, Piedmont. In England, more than a millennium earlier, it had been Wessex.

The ascent of Wessex came late. The first Anglo-Saxon kingdom to lift itself above its neighbors was Northumbria, whose heartland was what is now the northeast of England, and which was the dominant power in the seventh century. Mercia, rooted in the West Midlands, dominated the eighth. Only when these two monarchies had been reduced by Viking depredation did Wessex, their southern neighbor, take on the leadership role that was eventually to unify the English-speaking peoples.

When the eighteenth- and nineteenth-century radicals defined nationhood, they thought largely in linguistic terms—though they allowed that language was not always the whole story. A single people might speak different tongues while retaining a strong sense of patriotism, as in Switzerland. Conversely, different national identities might coexist within the same linguistic continuum, as among the various speakers of Serbo-Croat. Language, though, was their usual starting point, since it was a clear demarcator.

Nation-states are sustained by their citizens' awareness that they have something in common that they don't share with other peoples. Most national movements, whether they agitate for the gathering of a people into a single state or for their independence from another state, are in practice defined with negative reference to a foreign identity. For the Italians, that foreign identity was Austrian. For the Germans, it was French.

Ninth-century Englishmen were no exception. Indeed, they had more cause than most to define themselves against the foreigners who had come among them, for those foreigners were the most feared and reviled people in Europe at that time: Vikings.

At the end of the eighth century, raiders from Scandinavia began to ravage England's eastern shore. In 793, they sacked the Holy Isle of Lindisfarne, no doubt wondering at their luck in happening upon a civilization that heaped gold and silver treasures in coastal monasteries whose guardians were forbidden by their religion to fight. Like the Angles and Saxons before them, they soon turned from pillage to invasion, seizing tracts of land in northern and eastern England.

As far as we can establish, the first English settlers—the Angles, Saxons, and Jutes—had been of similar stock, speaking cognate dialects and following shared customs. But three hundred years of separate development had sundered them from the Germanic peoples on the other side of the North Sea. The Anglo-Saxons had converted to Christianity; the Danes had remained pagan. The two peoples spoke related languages but would no longer have been able to understand each other in ordinary conversation. Here, in short, was an alien presence that made the English acutely aware of what they had in common.

One by one, the Northmen conquered the Anglo-Saxon kingdoms. They swallowed up East Anglia. They overran Northumbria, putting a puppet ruler on its throne. They seized half of Mercia, leaving the other half to throw in its lot with Wessex.

They came close to taking Wessex, too, but were eventually defeated by the only English monarch ever to have been named "the Great": King Alfred. A pious and thoughtful man, who dreamed of making his throne a focus of learning, Alfred came close to losing it. Early in his reign, caught by a surprise Danish attack, he was forced to flee to the marshes of Somerset, desperate and deserted. It was at this time, according to a later chronicler, that he was supposed to have burned some griddle cakes that a swineherd's wife, unaware of his royal identity, had asked him to keep an eye on. The story is apocryphal: berated by the angry wife, Alfred is said to have stopped feeling sorry for himself and remembered his duty to his countrymen. What is not apocryphal is that, from that moment, he began winning battles. The Danish advance was rolled back.

The reconquest of England was not a short or smooth process. Alfred's modest territorial gains were built upon by his descendants. His grandson, Athelstan, established his sway over a territory roughly congruent to modern England, but wars between Anglo-Saxon and Danish kings were to continue right up to the Norman Conquest. Nonetheless, the birth of England as a nation-state can be dated to Alfred's wars. In 876, according to the Anglo-Saxon Chronicle, "all the English people who were free to give him their allegiance [in other words, were not under Danish occupation] owned Alfred as their King."

This is not the first reference to the English people. The concept of an English race, an Angelcynn, had existed from at least the eighth century, possibly earlier. What was new was the idea that all the Angelcynn, by virtue of their common identity, should recognize a single sovereign.

We shouldn't overstate our case. Victorian historians sometimes wrote teleologically, assuming that the emergence of an English state was inevitable, and viewing the absorption of the Danes into it as part of an inexorable historical process. But England in the ninth and tenth centuries was also part of a looser Nordic community. Between 1018 and 1035, England had a Danish king, Cnut, who also made himself ruler of Norway and part of Sweden. Indeed, given the respective size of the English, Danish, and Norwegian populations, we can speculate that, had William the Conqueror not wrenched it out of the Scandinavian world, England might have become the dominant Nordic power. There is evidence, for example, that Danish bishops were acknowledging the authority of the Archbishop of Canterbury, and English currency was common tender in eleventh-century Denmark.

Nonetheless, the concept of Englishness, and of an English state, remained strong from Alfred's time onward. Strong enough, indeed, to absorb the descendants of the Danes who had arrived as conquerors.

When Alfred drove the Vikings out of Wessex, the eventual peace treaty allowed for a large part of northern and eastern England to retain a Danish system of law and government, hence its name: the Danelaw. The Danelaw, with its capital at York, would last for seven decades before eventually joining the English kingdom (its inhabitants, of course, would not have been wholly of Danish descent, but would have included many Anglo-Saxons).

Our language altered in interesting ways during the period of

Anglo-Danish cohabitation. Although the two tongues were largely mutually incomprehensible, they had a common root, and their speakers might make themselves understood by speaking a simplified pidgin. Old English and Old Norse often had similar basic words, but their suffixes and prefixes rendered communication almost impossible. It was only by dropping the different ways to conjugate a verb, the different forms of irregular plural, the different noun cases, that Saxons and Danes could make themselves readily understood.

It was during this period, in the Danelaw, that English began its move toward its modern form. Compared with most other languages, English is uninflected. We give meaning to a word by putting other words around it, not by altering its form. In most of our verbs, only the third-person singular is different (I eat, you eat, he *eats*, we eat, you eat, they eat). Many irregular plural noun forms that existed in Old English were swept away during this period, to be replaced with the simple rule, in most cases, that you add an *s*. All these changes, along with a substantial expansion of our vocabulary through the annexation of Viking words, date from the absorption of the Danes.

With this linguistic assimilation came an assimilation of outlook and identity. The descendants of the Vikings were soon looking to English kings to defend them from later waves of Nordic sea raiders. Within a century, they were speaking English and self-identifying as Angelcynn.

It's not hard to see why. The tenth century was the high point of the early English state: law-governed, unified, and wealthy, it began to develop several traits that set it apart from the neighboring landmass. The characteristics that led, in time, to parliamentary democracy, independent justice, and personal freedom were already there, pulsing in the womb. Let us now consider the most important of those characteristics: the rule of law.

THE LAW OF THE LAND

The first thing that strikes the historian, as he surveys the written records of the Anglo-Saxons, is that they were a litigious people. A great deal of their time was taken up with writing legal codes, some of them in the form of royal decrees, others as restatements of inherited, traditional laws.

They might have arrived as violent raiders but, within a remarkably short time, they formed the habit of seeking to settle their quarrels peacefully, by laying them before a court. That they did so tells us a great deal about the homogeneity and orderliness of early English society.

The cultural and linguistic unity of England was reflected in a universal legal system. Lawlessness tends to multiply when the government's authority is weak or disputed, and England was the most stable and united nation-state in Europe. On the Continent, seigneurial justice was common. The great magnates *were* the law on their own estates. But England, before the Normans came, had no feudal aristocracy. It had its great men, as all warrior societies had, and many of them held large estates, though these tended to be fragmented across many counties. But the great men never constituted a hereditary caste with legal privileges—such as, to cite the most notorious example from Europe, exemption from taxation. They were as subject to the law of the land as anyone else.

We can see, in its genesis, the extraordinary idea of equality before the law. James Campbell cites the case of a letter from Queen Edith, the widow of Edward the Confessor, written to the court of the hundred of Wedmore in Somerset, and asking for "a just ruling concerning Wudumann, to whom I entrusted my horses, and who has for six years withheld my rent." We don't know which way the Wedmore court decided. But it is hard to imagine a similar case anywhere else in the world in the eleventh century. Here is the greatest lady in the kingdom petitioning a minor public court for justice in a personal matter.

It wasn't until the mid-twelfth century, in the reign of Henry II, that the term "common law" was used. But the concept was Anglo-Saxon. A wealth of recent research, above all by the foremost historian of the period, Patrick Wormald, has shown that the laws encoded by the Normans largely predated the Conquest, and that the principles and practice of common law survived almost unchanged at the level of the county and its chief subdivision, the hundred.

Almost a third of the human race now lives, wholly or partly, under a common-law system. Along with the English language, the common law is the main unifier of the Anglosphere. It applies in most former British territories—though not in Quebec nor yet, curiously enough, in Scotland. It is used, too, in Israel. A variant of it, which grew up alongside its English cousin from common ancestry, can be found in Scandinavia.

What distinguishes the common law from the Roman law that predominates in Continental Europe and its colonial offshoots? Chiefly this. The Continental legal model is deductive. A law is written down from first principles, and then those principles are applied to a particular case. Common law, to the astonishment of those raised in the Roman or Napoleonic systems, does the reverse. It builds up, case by case, with each decision serving as the starting point for the next dispute. It applies a doctrine known to lawyers as *stare decisis*: previous judgments should stand unaltered, serving as precedent. Common law is thus empirical rather than conceptual: it concerns itself with actual judgments that have been handed down in real cases, and then asks whether they need to be modified in the light of different circumstances in a new case.

It is therefore sometimes known as "judge-made law," but, as the philosopher Roger Scruton, who himself trained as a barrister, points out, "the common law is no more *made* by the judge than the moral law is *made* by the casuist." It is more useful to think of the law being *discovered* in stages. Just as a good man is not necessarily a skilled philosopher, so the common law recognizes that doing the right thing is not necessarily the same as explaining the principles that make it right. We often know what is the correct way to behave without being able to put our reasons into words. The same is true of legal disputes. An individual case might have an obviously just remedy, one that conforms to everyone's idea of fairness, and yet whose resolution doesn't translate neatly into a general principle. The pragmatic nature of the Anglosphere peoples, their dislike of purely theoretical reasoning, was built from the first into the way they made—or, rather, discovered—their laws. The law didn't real-ize an abstract principle; rather, the principle was pieced together in stages from actual rulings.

The notion of "the law of the land"—of a law, that is, that was the property of all, not a device of the ruler—could be seen in the nature of the criminal justice system. Courts were dispersed and local. The more important cases were tried at county level, the lesser at the level of the smaller unit, the hundred. The most important cases of all were referred up to national level, which eventually gave rise to the medieval development of a high court, whose judges would tour the county courts in turn (in a "circuit"). Highest of all was the Anglo-Saxon Witan, the supreme council that would adju-dicate the cases of national importance.

At an early stage, the most important cases began to be tried in courts that included a selection of ordinary citizens: a jury. It's true that Anglo-Saxon juries were very different from the ones that meet in Anglosphere countries today. Their role was as much to weigh the character of the accused as to make a judgment on the basis of evidence. Nonetheless, they served to domesticate the law, emphasizing that it was there for the whole country, and was not simply a tool of government. As Tocqueville, who saw the common law as the chief guarantor of Anglo-American freedom, put it, "The jury is the most direct application of the principle of the sovereignty of the people."

The jury system, while flawed, as all human mechanisms must be, is yet a pillar of Anglosphere liberty. It ensures that questions of fact are distinguished from questions of law, that the assumption of innocence is no mere formality, and that the prosecution must establish its case beyond reasonable doubt. It also prevents the law from straying too far from the commonsense prejudices of the population, since juries refuse to convict when a crime could result in a disproportionate penalty. Above all, it involves the entire nation in the administration of justice. Jury service was—and remains—an obligation. The law thus rests, in practice as well as in theory, on every household in the country. Truly, it is the law of the land.

Being the law of the land, rather than of the King, Anglo-Saxon common law had four further properties that have served, to this day, to distinguish it from most civil law systems. First, it laid particular emphasis on private ownership and free contract. There was no sense, as there was and still is in many European countries, that tenure was the state's to determine, and that property rights were contingent. The law came up from the people, not down from the government.

Even now, most European legal systems limit an individual's liberty to dispose as he pleases of his goods. When he dies, for example, a proportion of his estate is generally reserved for surviving family members. England was exceptional in elevating the wishes of the individual, even when dead, above the perceived need of the surviving community—a peculiarity that, as we shall see, had huge consequences, making possible the trusts and foundations that are the basis of civic society, and creating an individualist, rather than a peasant, society in the countryside.

Second, common law is based on the notion that anything not expressly prohibited is legal. There is no need to get the permission of the authorities for a new initiative. Again, even now, we see this consequence of the difference between British and Continental practice. British Euro-skepticism owes a great deal to a resentment of what is seen as unnecessary meddling, but, to the Eurocrat, "unregulated" is more or less synonymous with "illegal." I see the difference almost every day. Why, I often find myself asking in the European Parliament, do we need a new EU directive on, let's say, herbal medicine? Because, comes the answer, there isn't one. In England, herbalists have been self-regulating since the reign of Henry VIII. In most of Europe, such a state of affairs could never have come about.

Third, the invigilation of the law of the land was everybody's business. The policeman was and is a citizen in uniform, not an agent of the state. He has no more legal powers than anyone else, except to the extent that those powers have been temporarily and contingently bestowed on him by a magistrate. In many parts of England, law officers were directly answerable to their communities— something that is now once again true of the entire country.

In November 2012, England introduced—or, rather, reintroduced— direct accountability for its law officers. Police forces were made accountable to elected county commissioners. I had had some role in promulgating the policy, and had originally wanted the commissioners to be called sheriffs, as in the United States. In a sad comment on how the English have lost their sense of history, the Home Office dropped that name when its focus groups told it that the title sounded "too American." In fact, the shrievalty is one of many institutions that are often thought to have American roots but originated in England.

Finally, and most important, the fact that the law was national rather than monarchical implied the need for an ultimate popular tribunal to determine it. That tribunal emerged in Anglo-Saxon times, reemerged after the calamity of the Norman invasion, and survives to this day as Parliament, which, on some formal occasions, still calls itself the High Court of Parliament.

The Anglo-Saxon Witan grew out of the lesser shire and hundred courts as the supreme council of the nation. The inferior assemblies, though they were conceived as courts, also had many of the characteristics of a parliament. They were gatherings not only of big landowners and clergy but also of all freeholders. In addition to legal disputes, they would often settle the distribution of the tax burden in their locality. They became, in effect, local councils: places where the affairs of the shire or hundred were publicly deliberated. As England became a united nation, these attributes were replicated at national level by the Witan, which in time reemerged under the Normans as the Parliament, which sits to this day.

It is often forgotten that Parliament began as, and in a few formal senses remains, a supreme court. The great constitutional upheavals in England, from Magna Carta in 1215 to the Glorious Revolution of 1688, were experienced as legal crises, and all were eventually settled by new legal charters. It is to the story of Parliament that we now turn.

THE WITAN OF THE ENGLISH PEOPLE

Two thousand years ago, the Roman historian Tacitus told his countrymen something quite extraordinary about the barbarians who lived beyond their frontiers. The primitive German tribes, he wrote, were in the habit of deciding their affairs through open-air clan meetings. Their chiefs were not autocrats, but governors by consent. Their rule rested upon *auctoritas* (the ability to inspire), rather than *potestas* (the power to compel). Their peoples were not subjects, but free and equal participants in the administration of their affairs.

Tacitus's *Germania* was hugely popular in the Anglosphere during the seventeenth and eighteenth centuries—as it was in Germany, Scandinavia, and, indeed, France. (Montesquieu quoted it frequently and approvingly, convinced that the French were of predominantly Frankish rather than Gallic origin, and thus Germanic, too.) The idea that free parliaments were a prehistoric birthright of the Teutonic peoples was naturally flattering to their descendants—especially those who were arguing for a more democratic form of government in their own times.

It was true that Tacitus, like most historians, had been writing with an agenda. By holding up the German savages as models of republican virtue, he hoped to shame his Roman fellow citizens out of what he saw as their authoritarianism and servility. But this, of course, only endeared him the more to British, American, and German writers, sustaining their sense that they were by inclination and temperament more suited to liberty than the Latin nations were.

During the twentieth century, such views became unfashionable. They were deemed out of joint with the times: jingoistic, anachronistic, even racist. And yet no one has ever been able convincingly to show that Tacitus was wrong. Tribal meetings do indeed seem to have been a common feature of very early Germanic society. Later sources take them for granted, as part of the immemorial way of doing business. While their composition and functions varied, certain core tasks—the confirmation of kings, the settlement of legal disputes, the allocation of taxation—crop up in many of the post-Roman Germanic kingdoms. It seems vanishingly unlikely that these proto-parliaments emerged through a process of parallel evolution rather than from a common tradition.

Certainly the sources show the early English kings ruling through councils of their people. In the 620s, for example, Edwin of Northumbria changed his religion following a meeting with the "wise men" of his realm—much as the Tudors were to do when they got Parliament to ratify the adoption of Protestantism. We read of Ine of Wessex issuing laws through councils of his bishops and nobles in the 690s. We know, too, that these councils, especially at county level, had a role in allocating taxation. One early Mercian source concerns an exemption from the share of the fiscal burden determined by a local *folcegemot* ("people's meeting").

It would be absurd, of course, to claim that these tribal councils were democratic legislatures. Nonetheless, as the Anglo-Saxon kingdoms developed and consolidated, we can see several features that these meetings of "wise men" (*witan*) had in common with later, recognizably parliamentary, assemblies.

First, they were the forum in which the sovereignty and legitimacy of the regime were affirmed. They were, for example, one of the rare occasions on which monarchs would physically wear their crowns and, later, carry scepters and other kingly regalia. The same remains true in Britain today.

Second, they were there to ratify the most important decisions in the kingdom: large grants of land, for example, and the resolution of major legal disputes.

Third, at least by the ninth century, they became not just frequent but regular, often holding their sessions to coincide with the major Christian festivals: Easter, Christmas, and Whitsun. (These dates still define the parliamentary calendar in many Anglosphere countries.)

Fourth, they brought together the great ecclesiastical and temporal lords as well as some of the second-order landowners. Bishops, ealdormen (county magnates), and thegns (lesser landowners) would meet jointly in session. This ancient composition can yet be glimpsed vestigially in the British House of Lords, which contains bishops as well as ninety-two hereditary peers who, in a very British compromise, survived the expulsion of the rest of their caste in 1998. Certainly the composition resembles that of later medieval assemblies, which were undeniably parliamentary in character. As J. R. Maddicott, the chief authority on the development of Parliament, says of the tenth-century Witan: "Substituting 'earls' for 'ealdormen' and 'barons' for 'thegns,' we are not so very far from the general look of an early parliament."

We can observe the formalization of the institution in the name. In the sources, meetings of *witan* ("wise men") become "the Witan." As early as the 880s, Alfred's council is described as "ealles Angel-cynnes Witan": "The Witan of all the English." In the later sources (though much more rarely), we read of something more apparently regular: "Witenagemot," "The Meeting of the Witan." It was this last variant that the Victorians, keen to establish the longest possible pedigree for the Parliament of their own day, favored. And, as usually happens, there was a backlash by a later generation of historians, who saw such pedigrees as smug and nationalistic. More recently, though, a sense of perspective has returned.

"That representative institutions have their roots in the dark-age and medieval past is not an anachronistic view; rather it is fully demonstrable," says James Campbell, professor of medieval history at Oxford and an unmatched authority on the late Anglo-Saxon state. "It does indeed look as if the history of constitutional liberty has important beginnings in Anglo-Saxon England."

J. R. Maddicott, whose *Origins of the English Parliament* is the standard work on the subject, contrasts the Witan to other European assemblies and concludes that it was qualitatively different in its more representative composition, in its powers of taxation, and, above all, in its survival. "In other parts of the West, the Germanic legislative tradition died out in the tenth century. Its energetic preservation and promotion in England was quite exceptional." The Witan was not only a partner in royal lawmaking; it was also a guardian of the established law, willing, on occasion, to lay down terms to the King. "We need not baulk," says Maddicott, "at the notion of English exceptionalism."

This exceptionalism is most clearly visible in the dynastic disputes of the early eleventh century, before the catastrophe of the Norman Conquest.

Early English kingship had always been partially elective: notions of legitimate succession and divine right came later. Before the Norman Conquest, the coronation would take place only after the succession had been determined by the Witan. Thus, for example, Alfred's son Edward was "chosen as King by the great men of the kingdom." Eadred became king in 946 "by the election of the chief men."

An element of reciprocity was established by the coronation oath, which makes its first appearance at King Edgar's consecration in Bath in the year 973. It was, and for a long time remained, the only part of the ceremony conducted in English rather than Latin, suggesting that it was intended to be understood by all present. The promises that King Edgar gave on that occasion were remarkably similar to the ones given by his descendant, Elizabeth II, 980 years later: to defend the land, uphold its laws, protect its church, and rule justly.

Here, in fetal form, is the idea of government by contract: the notion that rulers and ruled are bound by a specific agreement and that, if the ruler breaks his part of the bargain, the deal is off. Contracts were and are enormously important in Anglosphere societies. They informed people's understanding of law, business, religion, and politics. They eventually became, above all through the writings of John Locke, the basis of the predominant Anglosphere theory of government. The state was held to be an expression of a primeval contract made in the remote past by our ancestors: what Locke called the "original compact." Locke's idea took substantial form in the U.S. Constitution and later spread throughout the free world.

The really critical innovation, though, was the notion that the contract should be enforced by a representative assembly. Plenty of European medieval kings took coronation oaths. The oath, like the moment of anointment, was religious in inspiration. It was a promise made to God in the presence of the country's chief bishop, who would carry out the most important part of the ceremony when he anointed the head of the new monarch with oil. This moment, the unction, remains at the center of the coronation ceremony. When the present queen was crowned sixty years ago, many of her subjects bought televisions so as to be able to watch the pageant. The only moment that was hidden from the cameras was the unction.

What was truly exceptional was not the monarch's oath, but the idea that he might be held to it by representatives of his people. The King wouldn't be the sole judge of whether his promise to uphold the laws had been kept. That task would fall to the Witan.

In 1014, something almost miraculous took place. England had been ruled by one of its most unfortunate kings, Æthelred, known to history as Æthelred the Unready. Unready is a literal transcription of his contemporary nickname, "Unred," meaning "Ill-Advised," for Æthelred ruled during a period of intense Danish attacks and repeatedly bought the invaders off with gold. This gold had to be levied from his people through heavy taxation—something that has never been popular in the Anglosphere. Worst of all, the bribes, known as Dane-geld, didn't work. The Danes would accept the payment and promise to go home, only to return the following campaign season and demand more. As Kipling was later to put it:

> And that is called paying the Dane-geld;
> But we've proved it again and again,
> That if once you have paid him the Dane-geld
> You never get rid of the Dane.

By 1014, a series of disasters had overtaken the English. The Danes had seized London, forcing Æthelred to flee into exile in Normandy. What happened next was, at that time, without precedent in the world. The Witan offered Æthelred the chance to return to his throne only if he agreed to abide by their conditions. Specifically, there were to be no more excessive taxes. The old laws—the first appearance of that English notion of "immemorial custom" or "the good old laws"—must be upheld. And the King must pledge to be guided by the counsel of the Witan in future.

It was a remarkable, though curiously unremarked, development. When tracing the story of constitutional liberty in the English-speaking world, specifically in the form of the assertion of representative government over monarchy, historians point to Magna Carta, to the victories of Simon de Montfort, to the English Civil War, to the Glorious Revolution, and, finally, to the American Revolution. Yet here, fully two centuries before Magna Carta, we find a foreshadowing of the Glorious Revolution of 1688: a king being invited conditionally to the throne. The law is deemed to be bigger than he is.

It might be argued, of course, that the events of 1014 were atypical. The kingdom was imperiled, and its monarch was in an exceptionally weak position. Much the same, though, could be said of all the milestones just mentioned. In any case, the truly extraordinary thing about the recall of Æthelred is that it was not a one-off, but the beginning of a new constitutional order.

We know this because, when Æthelred died two years later, the English throne was offered to the Danish king Cnut on the same conditional terms. Cnut undertook, apparently willingly enough, to refrain from excessive taxation, to punish abuses by his officers, the reeves, and to uphold the law of the land.

What is perhaps most remarkable of all is that, in these compacts, the Witan was seen to have a representative function: it spoke, not just for its members, the thegns and ealdormen, but for the nation as a whole. The invitation to Æthelred to return from exile came from "all the Witan," but his reply was addressed to "all the people." He was, in other words, speaking through the assembly to the nation it represented.

Likewise, when Cnut was conditionally offered the throne by the Witan two years later, his acceptance took the form of a promise to uphold his side of the bargain. It was written in the form of a letter, in the English language and probably intended to be read out at shire courts, addressed to "his archbishops and diocesan bishops and Earl Thorkel and all his earls, and all his people, whether men of a twelve-hundred wergild or of a two-hundred, ecclesiastic and lay, in England, with friendship." The concept of what later generations were to call "virtual representation" can already be glimpsed.

When Cnut himself died in 1035, there was a full meeting of the Witan ("ealre witena gemot") at Oxford to determine who should rule next. It is no longer possible to see such meetings as an exceptional response to a disputed succession: they had become the norm.

Thus, in 1041, when Edward the Confessor was recalled from Normandy and offered the throne, he was met at Hurst Head on the Hampshire coast by "the thegns of all England" and told that he might be awarded the crown if he swore an oath to uphold the laws of the time of King Cnut—which had already, in the minds of Englishmen, become "the good old law" to which their descendants would regularly demand that their sovereigns conform.

Ponder for a moment the astonishing nature of that development. Here is a king being met on a sandy spit of land on the Solent and presented, not with a set of arbitrary demands, but with an already familiar—we might almost say constitutionally regular— requirement that he accept the law as defined by the body that speaks for the nation.

England was, by contemporary European standards, a prosperous enough place. But these were nonetheless wretched times for all humanity. Life expectancy in England, for those who survived infancy, was around 42: a study of sixty-five burials from between AD 400 and 1000 found no one who had lived past forty-five. Literacy was confined to a minuscule number of monks and scribes. Medical science was unbelievably primitive. Life for almost the entire population consisted of backbreaking labor seven days a week. The entire population of the British Isles was no more than two million. Yet we can discern something recognizable as modern constitutional government. The King is not above the law (*rex lex*), but rather the law is above the King (*lex rex*). The interpretation of that law is not left to the sovereign's conscience, nor to the clergy, but is in the hands of a representative body that self-consciously speaks for the nation as a whole.

More than seven centuries later, drawing up the Massachusetts Constitution in 1780, John Adams was to come up with one of the finest and simplest definitions of constitutional government. The powers of the commonwealth were to be divided and balanced, he wrote, "to the end it may be a government of laws, not of men." Though they would not have put it in such terms, the earliest speakers of his tongue plainly had some notion of what he was driving at.

Tenth-century England had undeniably started down the track to constitutional liberty. What might have happened had it continued on that path we'll never know, because, in 1066, it was brutally wrenched out of the Nordic world and subjected to European feudalism. Harold Godwinson, an English nobleman with scant claim to the throne, but with the unequivocal backing of the Witan, was deposed by William of Normandy, who had his own ideas about the duties owed to a king. It was a calamitous defeat for England, for the Witan, and for the development of liberty. Indeed, the next six centuries can be seen in one sense—and *were* seen by many of the key protagonists—as an attempt to reverse the disaster of 1066.

CHAPTER 3

Rediscovering England

"My son," said the Norman Baron, "I am dying, and you will be heir
To all the broad acres in England that William gave me for my share
When we conquered the Saxon at Hastings, and a nice little handful it is.
But before you go over to rule it I want you to understand this:

"The Saxon is not like us Normans. His manners are not so polite.
But he never means anything serious till he talks about justice and right.
When he stands like an ox in the furrow with his sullen set eyes on your own,
And grumbles, 'This isn't fair dealing,' my son, leave the Saxon alone."
—RUDYARD KIPLING, 1911

The northern nations had no idea, that any man, trained up to honor, and enured to arms, was ever to be governed, without his own consent, by the absolute will of another; or that the administration of justice was ever to be exercised by the private opinion of a magistrate, without the concurrence of some other persons, whose interests might induce them to check his arbitrary and iniquitous decisions. The King, therefore, when he found it necessary to demand any service of his barons or chief tenants, beyond what was due by their tenures, was obliged to assemble them, in order to obtain their CONSENT: And when it was necessary to determine any controversy, which might arise among the barons themselves, the question must be discussed in their presence, and be decided according to their opinion or ADVICE. In these two circumstances of consent and advice, consisted chiefly the civil services of the ancient barons; and these implied all the considerable incidents of government.

—DAVID HUME, 1778

BASTARD NORMANS

On July 3, 1940, Admiral Sir James Somerville issued the saddest order of his career. France had been occupied by the Nazis and was required under the armistice terms to transfer its Mediterranean fleet to German command. The British couldn't allow such a development: Italy had entered the war on Hitler's side, and control of the Mediterranean was at stake.

Winston Churchill ordered a larger British force to confront the French fleet off the Algerian naval base of Oran. The French admiral, Marcel-Bruno Gensoul, was given three options: to take his ships to British waters and carry on the struggle; to remove them from the theater of operations and keep them in the West Indies for the duration of the war; or to scuttle them.

All three options were turned down and, as the sultry day wore on, a final ultimatum was issued and rejected. At last, Admiral Somerville ordered his ships to shell the French fleet, the only occasion the British and French navies have exchanged hostile fire since Trafalgar. For ten minutes, great geysers of water shot into the sky, soon joined by black smoke from the battleship *Bretagne*, which was badly hit. No fewer than 1,297 Frenchmen were killed and 351 injured, by far the worst naval losses suffered by France during the war. There were no British casualties.

Somerville was sickened by what he later called "the most unnatural and painful decision" of his life. He passed a grim and silent evening in the mess, where many of his officers had tears in their eyes. But he couldn't help noticing that, on the lower decks, a very different attitude prevailed, most sailors cheerfully declaring that they "never 'ad no use for them French bastards."

It was an extreme illustration of an age-old social divide. The English (and later British) upper classes tended to be Francophone and Francophile. Yet theirs was a minority tendency, one that opened them down the centuries to accusations of being effete and unpatriotic.

That class division can be traced right back to the Norman Conquest, which placed England under a French-speaking aristocracy. It was to be more than three centuries before English again became the language of Parliament, the law courts, the monarchy, and the episcopacy. Certain parliamentary procedures are still, a millennium after the Conquest, conducted in Norman-French. The Queen's approval of legislative bills, for example, is announced with the phrase "La Reine le veult."

The native English, disinherited and resentful, projected their resentment onto French-speakers in general. The popular stereotype of the Frenchman closely resembled the radicals' stereotype of the aristocrat: mincing, epicene, sly.

Even today, most Britons suspect (with good reason) that their elites are more Europhile in general, and more Francophile in particular, than the country at large. By "Europhile," they don't simply mean readier to accept EU jurisdiction, though that belief is demonstrably accurate. "Europhile" has wider connotations: of snobbery, of contempt for majority opinion, of the smugness of a remote political caste.

The extraordinary thing is that we can find no period in the past nine hundred years when such a sense was absent. The linkage between French manners and upper-class decadence has been made in England (then Britain, then the Anglosphere as a whole) by every generation.

The political broadcasts that attacked John Kerry and Mitt Romney because they could speak French echoed the diatribes of Johnson and Hogarth against the effeminate aristocrats of their own day, whom they accused of picking up nasty foreign habits when they went on the Grand Tour as young men, and who consequently tended to patronize Continental rather than native artists. "Painted faces, fashions, frippery and foppery" was how the eighteenth-century Scottish author Tobias Smollett summarized French civilization.

The eighteenth-century patriots were in turn echoing the seventeenth-century parliamentarians, who had raged against the Francophile tastes of the Stuarts, and who explicitly defined their mission as "throwing off the Norman yoke."

Go back a little further and see how Shakespeare, in *Henry V*, contrasts the rude, brave, quarrelsome English soldiers with the mannered French knights. ("Bastard Normans! Norman bastards!" yells the Duke of Bourbon as he realizes that the English advance has not been checked.)

Reach back further still and see how Chaucer subtly mocks the Anglo-Norman aristocracy, who, though they think themselves so grand, are in fact cleaving to an antique and incorrect form of French. He laughs softly at the Abbess in the Canterbury Tales:

And French she spake, full fair and featously,
After the school of Stratford atte Bowe;
For the French of Paris was to her unknow.

Chaucer's works were revolutionary not only in their dramatic qualities but also in the fact of being written in English. In his life-time, most literature was still aimed at the upper classes and addressed to them in French. English was a rustic vernacular. Only in Scotland, whose court adopted English around the time of the Conquest, was the language widely used for literary purposes.

Chaucer, like so many writers after him, was unabashedly patriotic about his national language:

Right is that English, English understand
That was born in England.

How extraordinary that such a thing should need to be written of the tongue spoken by 99 percent of Englishmen fully 320 years after the Conquest.

Go any further back and you reach the period when the Anglo-Saxons were conscious of being a subjugated race, governed by a foreign caste and in an alien tongue.

The last direct Anglo-Saxon rising against William the Conqueror had been put down in 1071, when Hereward the Wake and his followers were defeated in the marshes of Cambridgeshire. But the inner resistance of the English people did not vanish into the swamp with Hereward.

A century after the Conquest, we come across a prophecy that was anecdotally attributed to Edward the Confessor, whose death had prompted William's invasion. It was said that when a green tree was cut down and moved three acres, and yet miraculously returned to its root to bear fruit, then the kingdom would be restored to the English. Some saw the story as an allegory for the accession in 1154 of Henry II, whose grandmother Margaret, Queen of Scotland, had been Edward's daughter. The King might be, to all outward appearances, a Frenchman, but he was nonetheless a descendant of Edward the Confessor, and thus of King Alfred. His rule, said the optimists, represented a restoration of sorts of the old dynasty: the regrafting of the green tree.

Others were more cynical. The story, they said, was a fable of hopelessness. It was as unlikely that a felled tree would return to its root as that an Englishman would ever be free in his own land.

As time passed, the Anglo-Saxon opposition ceased to be linguistic and took on a class character. With their native aristocracy
slain, exiled or expropriated, most English-speakers were reduced to
landless poverty. Their national struggle became a rebellion against
the rule of an alien elite.

The hostile march on London by farmworkers and artisans in
1381 is known to historians as the Peasants' Revolt. But the word
peasant didn't exist in England at that time—largely because, as we
shall see, the thing itself didn't exist. The demands of the protestors
harked back to the old order. Their leader, an artisan named Wat
Tyler, demanded to meet the king, Richard II, who was then fourteen. To the horror of his counselors, the young sovereign agreed
to ride out to Smithfield for a parley with the ragged rebels. Tyler
asked the king to end the system of serfdom that the Normans had
introduced. Specifically, according to the contemporary chronicler
Henry Knighton, he wanted "the right of buying and selling freely
in towns and out of towns, and the right of hunting in all forests,
parks and commons, and of fishing in all waters, which men of the
English race lost at the Conquest."

Writing in 1825, the French historian Augustin Thierry declared
that "the great insurrection of 1381 would seem the last term of a
series of Saxon revolts." Nineteenth-century historians were keenly
interested in the national identity of their subjects. Indeed, until very
recently, it was normal to see history in broadly ethnic terms. Hence,
for example, Catherine the Great's pithy verdict on the French Revolution: "The Gauls are driving out the Franks." Such judgments
are now, of course, deeply unfashionable, which can blind us to how
much they mattered at the time.

English identity after the Conquest remained a powerful and
enduring phenomenon; more powerful, certainly, than Anglo-
Norman identity, which is why the one eventually subsumed the
other. English national consciousness was bound up, at the time,

with precisely what later historians were to claim: a belief that freedom and equality before the law were natural, and that feudalism and serfdom were alien.

Is it fanciful to imagine a race-memory that stretches across a millennium? Shouldn't we rather see the embrace of Anglo-Saxonism by later generations, and the positing of a national identity in opposition to Frenchness, as an invention? Up to a point. Of course later writers pressed the past into their own narrative, as every generation does, consciously or not. Yet there is also such a thing as folk memory. Traditions can be passed down orally as well as in writing, in families as well as in schools.

Let me return to the analogy with my native Peru. Pizarro and his followers subjugated Peru every bit as thoroughly as William and his henchmen had subjugated England. Indeed, more thoroughly, because the technological disparity was greater. Most Peruvians lost their religion and their language as well as their land; but not their sense of nationhood, nor their memory of having been dispossessed. That memory informed the violence of the Shining Path and other terrorist movements in the 1970s and 1980s, four and a half centuries after the Spanish conquest—the same time span that separated William I from Sir Edward Coke and John Hampden, who popularized the "Norman yoke" rhetoric.

Peru is a unitary state, as Norman England was. All its citizens are notionally equal before the law, all adults vote, and there have been two indigenous presidents. Yet the racial divisions introduced at the time of the conquest are a constant and obvious presence, and a man's physiognomy remains a pretty reliable guide to his wealth and status. People don't need history books to tell them why.

The same was true at an equivalent date in English history: political power in the seventeenth century was still largely concentrated in the hands of descendants of the Norman invaders. Even today, nearly a millennium after the Battle of Hastings, the surnames of the soldiers who crossed the English Channel with William have an aristocratic timbre: Balliol, Baskerville, Darcy, Glanville, Lacy. And with reason. A study of family names between 1861 and 2011 showed that bearers of Norman surnames were, on average, slightly more than 10 percent wealthier than the mean.

Those cheerful sailors on Admiral Somerville's flagship took certain things for granted without stopping to examine them. It would not have occurred to them to think of their commander as anything other than a dutiful officer. They would not have dwelt—any more than he would—on the fact that his ancestor, Sir Gaultier de Somerville, had come over with the Conqueror, and had been granted vast estates in northern England at the expense of its native owners. It would simply have been assumed that, being upper class, the admiral was likely to be more sympathetic to the French than the men who served under him. That was how the world was.

BREAKING THE SHIELD-WALL

What do the following people have in common? George Bush, Barack Obama, Bill Gates, Justin Timberlake, Rupert Everett, Maggie Gyllenhaal, George Washington, and the Prince of Wales. The answer is that they are all descended from William the Conqueror. Around 90 percent of people with English ancestry and 60 percent of Americans are thought to carry the blood of that fecund Norman dynast—though, of course, most cannot trace their family trees back so far. The warriors and mercenaries who crossed the Channel with their duke were few in number. But they imposed their will on the defeated land as thoroughly as they imposed their seed.

The Norman Conquest was a tragedy for the indigenous population; and, like all the best tragedies, it might easily have played out differently.

Edward the Confessor, the last king of Alfred's house, died in January 1066 leaving no sons. His closest male relative, his grand-nephew Edgar the Ætheling, was considered too young to assume the throne. Edward was said by some to have promised the crown to his second cousin, William, Duke of Normandy. He was said by others to have revoked that promise on his deathbed, instead be-queathing the realm to his brother-in-law, Harold Godwinson, a powerful English nobleman. William's supporters maintained that Harold had promised to back their candidate when he had been shipwrecked in Normandy.

The truth is that neither William nor Harold had a blood claim to the throne. But, with a symbolism that Victorian historians were to find irresistible, William was backed by the pope while Harold was backed by England's Witan.

Harold's coronation, with the support of the great men of the king-dom, prompted William to prepare an invasion force. Landless nobles, younger sons, soldiers of fortune, and other men of violence flocked to his standard from all over Europe. Though the core of his force was Norman, knights came from Flanders, Brittany, France proper, and even Italy, drawn by the lure of plunder and confiscated land.

The Normans were Vikings who had conquered northern France in the tenth century: part of the demographic explosion that saw Northmen seize territory in England, Scotland, Ireland, Green-land, North America, Russia, Sicily, and even Asia Minor. Only re-cently had Normans adopted the French language: *skalds*, Viking poets, were reciting Norse sagas in the dukedom as late as 1028.

The Normans were a bellicose people whose prowess rested partly on their courage and partly on their adoption of cutting-edge military technology in the form of the armored knight. For once, the phrase "cutting-edge technology" is precisely apt: a Norman cavalry charge was an almost unstoppable force, the sheer weight of mailed men and horses smashing through the defenders' lines.

Normans also knew how to make use of a mixed force, supporting their knights with infantry, archers, and crossbowmen.

Yet, despite these advantages, it was only by the merest chance that William prevailed. Harold ruled the most united and affluent state in Europe, and could call on a formidable reservist army, the fyrd. As he prepared to meet William's invasion, he had cause to be confident. But his men were uneasy. They had seen ominous signs in the heavens—Halley's Comet passed England in 1066—and fretted that a series of misfortunes was about to overcome them. They were right.

First, Harold's brother Tostig, backed by the Norwegian king, Harald Hardråda, landed in the north, forcing the English monarch to wheel his army about and meet the new threat. Then the English fleet, which had been patrolling the Channel in wait for the Normans, found itself storm-damaged and running short of victuals, and put into port to take on more supplies.

It was at this moment that William crossed the Channel. Then, as now, an opposed amphibious landing was every general's nightmare. The Norman ships, laden with heavy cavalry, would have been especially vulnerable to attack. But, in the event, William landed unopposed in Pevensey in Sussex on September 28, 1066, while the English fleet was in Kent and the army in Yorkshire, triumphant but exhausted after defeating Tostig and the Norwegians.

Hurrying back with his weakened force, Harold met William's knights at Senlac Hill, near the Sussex town of Hastings. Even now, he might have won. One of the most effective tactics available to the Anglo-Saxon fyrd was to form a shield-wall. Harold ordered each man to link his kite-shaped shield with that of his neighbor, making an impenetrable line, bristling with swords and axes. Faced with such a mass, even the highly trained Norman horses balked. But luck was still with the invaders. When the Breton division on William's left fell back, the Anglo-Saxon fyrdmen broke ranks to pursue it. The shield-wall was fragmented, and William saw his chance. Harold's brothers, Leofwine and Gyrthe, were cut down in the Norman counterattack and Harold himself later fell—pierced through the eye, according to legend, by an arrow. Leaderless and demoralized, the Anglo-Saxons fled.

The deaths of the three Godwinson brothers robbed the English of an adult rival claimant, and William began to march through Sussex and Kent, receiving the submission of what was left of the southern English nobility and episcopacy. As he approached London, the authorities there agreed to recognize him if he would spare the city, and he was duly crowned King William I of England at Westminster Abbey on Christmas Day 1066.

The Norman Conquest was a cataclysm for the English people. The native aristocracy was dispossessed and banished. Some fled to Scotland, others to Ireland, still others to Europe. Many became mercenaries in the service of the Byzantine emperor, forming an elite unit known as the Varangian Guard. Some sources talk of these exiles founding a settlement on the shores of the Black Sea called New England.

Old England, in the meantime, was clenched in a mailed fist. Castles, until then a rarity, were built across the country. Many still stand, handsome and crenellated, their geographical situation advertising their grim purpose. For these were not defenses against a foreign foe, but instruments of internal repression, from whose arrow slits new proprietors peered out at a beaten people.

Having put down a northern rising with terrifying savagery, leaving stretches of country depopulated, William began to rule as an absolute monarch. Quite how absolute can be seen from his compilation, some years afterward, of a comprehensive inventory of the kingdom he now possessed. For there was no question in his mind that England was now his to do with as he would. As the Anglo-Saxon Chronicle put it, "So very thoroughly did he have the inquiry carried out that there was not a single hide of land, not even—it is shameful to record it, but it did not seem shameful to him to do it—not even one ox, nor one cow, nor one pig which escaped notice in his survey."

The resulting tome became known as the Domesday Book—*domesday* being the Old English word for the "day of judgment"—for it was said that you might as well try to deceive God when you came before Him as hide your goods or chattels from the King's officers. In the pages of that volume, published twenty years after the Conquest, we see how thoroughly the native English had been expropriated.

William had parceled out almost the entire country among his mercenaries and liegemen. At least 92 percent of England was owned by men who had been born on the other side of the Channel. Of more than two hundred landowners who held their estates directly from the King, only two were Anglo-Saxons: Thorkell of Arden and Colswein of Lincoln.

Outside a handful of towns, most English people were subject to the lord of their local manor. Equality before the law was forgotten as the new aristocracy settled down to enjoy its privileges. Men were required by law to work on their lord's estates, and forbidden to leave without his permission. In the wide spaces between towns—perhaps 8 percent of the population lived in settlements of 450 or more—Continental-style serfdom was imposed.

Englishness became, almost by definition, a badge of poverty and subjugation. Human nature being what it is, some Anglo-Saxons, especially London merchants, sought to shed their English characteristics in pursuit of social advancement, Frenchifying their speech and addressing each other as "sire," to the disgust of their Norman overlords.

Nor was social climbing confined to the towns. On one English farm in 1114, records the historian Peter Ackroyd, the workers were listed as being called Soen, Rainald, Ailwin, Lemar, Godwin,

Ordric, Alric, Saroi, Ulviet, and Ulfac. By the end of the century all those names had disappeared.

In a society where being called Roger or Robert or Richard marked you out as likely to be powerful and wealthy, it was no surprise that Norman names came into fashion among the Anglo-Saxons. One boy in twelfth-century Whitby changed his name from Tostig to William because he was being bullied.

Of the old English names, only five survived: Alfred, Edgar, Edwin, Edmund, and—the only one to remain truly popular, largely because of the cult of the old king, whom the Normans also claimed as their own—Edward.

The status of the defeated English is often illustrated with reference to the vocabulary of meat. The Anglophone farmer in the field used plain Saxon words for his livestock: *cow*, *pig*, *sheep*. But by the time these animals found their way onto his Norman master's plate, they had acquired French-derived names: *beef*, *pork*, *mutton*.

More telling, though, is the political vocabulary introduced under the Normans. Out go *witan*, *folkmoot*, and *folkright*. In come *fealty* and *homage*, *fief* and *vassal*, *villein* and *serf*. The progress toward personal liberty, free contract, and equal access to the common law was suspended. In the words of the early-twelfth-century chronicler Orderic Vitalis: "The English groaned aloud for their lost liberty and plotted ceaselessly to find some way of shaking off a yoke that was so intolerable and unaccustomed."

The notion of throwing off the Norman yoke would animate a later generation of Englishmen in their struggles with the Stuarts, and would later still be revived by American Patriots who raised English liberties to their highest and most sublime form.

Though we think of those struggles as progressive, their contemporary champions saw themselves as conservatives. In their own minds, they wanted to reinstate the settlement that they believed had existed before 1066. When they used the word *revolution*, they meant it in the sense of a full turn of the wheel, a restoration of that which had been placed the wrong way up. And, extraordinarily, they had a point: the rights and freedoms that they secured truly did have pre-Norman roots.

THE ROOTS OF OUR RIGHTS

A snapshot of England at the end of the eleventh century would have revealed a country under military occupation. The knights and sell-swords who had come over with the Conqueror knew that their rule depended on military technology. They threw up motte-and-bailey castles: swiftly built, defensible strongholds of earth, rock, and wood, which were later upgraded to the majestic stone keeps that can still be seen today. Like all occupying garrisons, the Normans relied on the active collaboration of a few and the passive acquiescence of many.

The flower of Anglo-Saxon England had been cut down, but the roots remained, stretching deep into the soggy earth. Ealdormen, thegns, and housecarls were broken as a class but, in the provinces, the most advanced administrative machine in Europe hummed on.

The old units of local government—shires, hundreds, wapentakes, parishes—survived largely unaltered and, in many cases, survive still. Shire courts carried on meeting as before, making fiscal allocations, adjudicating local disputes. They were not much affected by the displacement of the Anglo-Saxon upper classes, for they had never been gatherings of magnates: men holding only one or two hides of land would attend, giving the shire councils a representative character unknown at the time on the Continent.

The mechanisms for collecting geld and other taxes continued more or less undisturbed. So did the system of "frankpledge," whereby the adult male population was sworn to keep the peace, and divided into units of ten, nine of whose members were pledged to produce the tenth before the law if required.

Although the Domesday Book reveals that Anglo-Saxon land-owners had been liquidated, plenty of the administrators and townsmen in its pages bear English names. In the shadow (often literally) of William's new castles, the urban aldermanic class went about its business, innocently engaged in making money.

None of these things should surprise us. The Normans and their European confederates numbered perhaps eight thousand after the Conquest. They could hardly govern a nation of more than a million people other than through its existing officials, from reeves to parish priests.

It is natural that contemporaries, like historians, should notice changes rather than continuities. The changes were well chronicled, dramatic, and often sanguinary; the continuities unremarked, local, and bureaucratic.

The Witan gave way to a council of Norman barons, whose chief function was to aggrandize and glorify the monarch. Any notion of an assembly speaking for the wider nation—let alone being able to approve kings or impose terms on them—was lost. But, under the surface, in shire and hundredal courts, the practical, case-by-case accretion of common law continued. Eventually, during the reign of Henry II, the common law was given a national character, and its elements—including jury trials—were recognized by the central state.

The Norman kings might have seen themselves as absolute sovereigns, entitled to dispose as they pleased of every square inch of land in the realm. But they couldn't extirpate the notion of the law as the property of the nation, the protector of the individual. Nor could they eliminate the idea of important decisions being taken at public meetings.

These subterranean trickles, these provincial rivulets, eventually flowed together to form a torrent that smashed the dam of royal absolutism. When, a century and a half after the Conquest, Normans and Saxons joined together to impose their will on King John, and later established a parliament to hold his son, Henry III, to account, they were not simply reviving memories of the Witan. Rather, they were forming a national assembly out of the older, local assemblies that had survived the Conquest.

Such a thing couldn't have happened while the Normans regarded themselves as a separate race. But, from early in the twelfth century, we see the great barons doing what almost all immigrant rulers do, and adopting the identity of their new land. (I won't labor the parallel with my native Peru any further: the same process can be seen in many ages and nations.)

This new self-identification in part reflected the fact of intermarriage and intermingling, in part the gradual loss of their lands in Normandy as families divided their English and Norman estates among their heirs, and in part their awareness that they had inherited a mature and sophisticated civilization. The historians among them might speak French, but they focused on the story of England, not of the Anglo-Norman empire. William of Malmesbury recounted the history of the island from the *adventus saxonum*. Gaimar's *L'estoire des Engleis* was, in large part, a translation of the Anglo-Saxon Chronicle into French verse. Even the church leaders, foreign almost to a man, quickly adopted the cults of Anglo-Saxon saints.

Part of this interest was bound to focus on the peculiar way in which the English ordered their affairs: the way they had domesticated their legal system, the sanctity of contract, and, not least, the way they were represented in a national council.

The association of Englishness with common law and representative government long predates Magna Carta. There is a text dating from around 1140 called "Laws of Edward the Confessor," which purports to be a collection of the old king's laws drawn up four years after the Conquest. While we can be pretty certain that no such legal codification took place, we can again glimpse the notion of "the good old law," or of "immemorial custom," which was so central to English politics. According to the author, William the Conqueror summoned learned men from around the country in 1070 to compile the statutes, along with twelve *electi* from each shire. This summons, claimed the writer, was a revival, not just of Anglo-Saxon practice, but of the ancient elected assemblies that had met annually in the time of King Arthur—who, in those days, was believed to have been a historical figure.

That such a thing could be imagined, however wistfully, tells us something about the way in which twelfth-century Englishmen thought of their country. It is sometimes said that later generations of historians, especially in the nineteenth century, romanticized the proto-democratic elements of early medieval England. Perhaps they did. But they were following an extremely old tradition.

In any case, the way in which a country romanticizes its past is itself instructive. "Getting its history wrong," observed Ernest Renan, "is part of being a nation." English exceptionalism was defined with reference, not to racial characteristics, military prowess, or island geography, but to law, liberty, and representative institutions.

The defining moment for the barons came in 1204, when the king of France annexed Normandy, absorbing the ducal lands into his crown estate. No longer could the English upper classes think of themselves as part of a cross-Channel aristocracy. Though their speech, their music, their verse, and their dress remained French in inspiration, their political orientation shifted.

As long as Normandy remained powerful and autonomous, it had been possible to think of it as the preponderant part of the Anglo-Norman realm. William and his sons had been in the habit of spending most of their time there and his great-grandson, Henry II, in a reign spanning thirty-four years and eleven months, spent fully twenty-one years and eight months—63 percent of his time—across the Channel.

After 1204, though, the Crown's Norman possessions were reduced to Calais (which remained in English hands until 1558) and the Channel Islands (which to this day recognize the sovereignty of the Queen as Duke—not Duchess—of Normandy). No longer were there two parts of the realm; rather, there was an English kingdom with some tiny offshore dependencies.

Cut adrift from the land of their ancestors, the Anglo-Norman oligarchs were thrown back upon the land of their birth. Their identification with the Anglo-Saxons was all the stronger because King John had surrounded himself with foreign favorites, particularly southern Frenchmen from the lands he had inherited from his mother. Speaking Occitan rather than Norman-French, these men were as alien to the magnates as they were to the lower classes. As so often, the presence of foreigners led to the natives recognizing a shared kinship. Among the barons who guaranteed Magna Carta, we find the surnames of lieutenants who came over with the Conqueror: Clare, Bigod, Mandeville, Vere, FitzWalter. Oppression at the hands of the King John, and resentment against his Angevin courtiers, had made Englishmen of them. Yet again, we see the civic rather than ethnic nature of Anglosphere values, which spread among any population given the right circumstances.

The conveniently obnoxious King John provided such circumstances. It is hard to think of a worse English monarch. The only contender who comes close is James II, who also united the country against him through that peculiar combination of stubbornness, petulance, and caprice that is the mark of a weak man. In both cases, the kings were providentially bad: had they been just a little more trustworthy, just a little less obstreperous, the happy constitutional developments that followed from their reigns might not have occurred.

John was an abominable human being. Short, sly, vain, duplicitous, and autocratic, he was widely suspected of having ordered the
murder of his nephew, Prince Arthur. When he died, the contemporary historian Matthew Paris wrote: "Foul as it is, hell itself is
defiled by the foulness of King John." His Christian name, though
for many succeeding centuries the most common boy's name in the
land, was never again borne by an English monarch.

John's reign (1199–1216) unleashed a series of calamities upon
England, of which the loss of Normandy was just the first. In 1209,
his quarrels with the church became so nasty that the pope placed
the entire country under an interdiction. No religious rites—not
even weddings and funerals at first—could be carried out in England. Two years later, defeated and excommunicated, the king
turned 180 degrees, placing both his kingdoms (England and Ireland, which his father had conquered) under papal sovereignty, and
leasing them back for one thousand marks a year.

Taxes, as we have already observed, have always been unpopular
in the Anglosphere, and these were taxes necessitated by the monarch's vanity and fickleness. When an Anglo-German force was
routed by the French in 1214, it was the last straw.

The barons, resentful of the king's arbitrary taxes and confiscations, angry at his rejection of their advice, and bitter about his
promotion of foreigners, took to the field against him. They did so
at precisely the moment when they were starting properly to self-
identify as English, and to recall that the English had traditionally
found ways to hold their rulers to account.

On June 15, 1215, in a field near Windsor, an event of truly planetary significance took place. For the first time, the idea that governments were subject to the law took written, contractual form.
The king put his seal to a document that, from that day to this,
has been seen as the foundational charter of Anglosphere liberty:
Magna Carta.

THE GREAT CHARTER

London in August 1647 was a tense and frightened city. The English Civil War had exhausted the nation and coarsened its people. Parliament had emerged victorious, but it was becoming clear that the real power in the land was the military force that had defeated King Charles I: the New Model Army. Its soldiers were advancing on the capital, unpaid and angry.

In a gesture to the troops, Parliament appointed their commander, Sir Thomas Fairfax, as constable of the Tower of London. The first act of the Roundhead general, on taking up his post, was an encouraging one. He called for the greatest treasure in the Tower to be brought before him. Not a crown, nor a scepter, but a desiccated piece of parchment carrying barely legible Latin writing.

"This is that which we have fought for," he said reverently, "and by God's help we must maintain."

Fairfax's attitude to Magna Carta was by no means unusual. In every generation, the English-speaking peoples have tended to treat that text as their Torah, the script that sets them apart. The eighteenth-century radical MP John Wilkes, writing while imprisoned in the same tower, called Magna Carta "the distinguishing characteristic of all Englishmen," under which title he included Americans, whose rights he strongly supported—though, eccentrically, not Scots, whom he regarded as absolutist and Tory by temperament. Lord Denning, the most celebrated of all twentieth-century English jurists, declared: "Magna Carta is the greatest constitutional document of all times—the foundation of the freedom of the individual against the arbitrary authority of the despot."

The few copies that survive from the thirteenth century are mostly housed in England's cathedrals, tended like the relics that were removed at the Reformation. One is on display in the Australian

Parliament in Canberra. Another hangs next to the Declaration of
Independence in the National Archives in Washington, D.C. If we
are looking for the Anglosphere's formative text, here it is.

So it can be quite a shock to read the contents for the first time,
and see quite how interested the authors were in the treatment of
Welsh hostages, the borrowing of money from Jews, and the placing
of fish traps in the Thames.

While Magna Carta remains on Britain's statute books, most of
its clauses were repealed in the nineteenth century, when hundreds
of archaic and medieval laws were tidied up. Only three of its thirty-
seven articles remain in force.

Why, then, is Magna Carta exalted wherever English is spoken?
What makes it different? For one thing, the three remaining clauses
on the British statute books are not insignificant: one guarantees the
freedom of the church, another the ancient liberties of the City of
London and of other towns and boroughs. The third, Article 29, is
the most important, for it is the basis of due process as we understand it today:

> No Freeman shall be taken or imprisoned, or be disseised [dispos-
> sessed] of his Freehold, or Liberties, or free Customs, or be out-
> lawed, or exiled, or any other wise destroyed; nor will We not pass
> upon him, nor condemn him, but by lawful judgment of his Peers,
> or by the Law of the land. We will sell to no man, we will not deny
> or defer to any man either Justice or Right.

Note, again, that phrase "the law of the land." What is the source
of that law? Obviously not the King, for he is here agreeing to bind
himself to it. Nor yet the bishops and barons who sealed the charter
alongside their sovereign: they, too, are pledging to abide by something bigger than they are. What the Great Charter enshrined, in

statutory form, was that the supreme power in the land was not the executive, but a set of fixed legal principles; and that, in any clash between the two, the law would prevail over the government.

As the eighteenth-century jurist William Blackstone put it in his elucidation of that clause:

> Since the law is in England the supreme arbiter of every man's life, liberty, and property, courts of justice must at all times be open to the subject, and the law be duly administered therein. The common law depends not upon the arbitrary will of any judge; but is permanent, fixed, and unchangeable, unless by authority of parliament.

Blackstone, as much as Locke, was the godfather of the American Revolution. His great work, *Commentaries on the Laws of England*, was said to have been the most widely owned book in the Thirteen Colonies after the Bible: every attorney kept a copy in his saddlebags. And, in truth, enthusiasm for Magna Carta was always stronger in North America than in Britain. (Then, as now, Americans tended to call it *the* Magna Carta.)

The site where the document was sealed, at Runnymede, lies in my constituency. It went unmarked until 1957, when a memorial was finally erected there by the American Bar Association. The spare inscription on that memorial is the answer to anyone who asks why we still make such a fuss of an eight-hundred-year-old treaty between a king and some aristocrats: "To commemorate Magna Carta, symbol of Freedom Under Law."

There was some amusement in 2012 when, interviewed on U.S. television by David Letterman, the British prime minister, David Cameron, was unable to render the words "Magna Carta"—the Latin for "Great Charter"—into English (Winston Churchill once translated a piece of Latin in the House of Commons "for the benefit of any Old Etonians who may be present"). Cameron's American audience turned out to be far more familiar with the document than he was. Not that this should surprise us. The United States, which emerged from the purest distillation of Anglosphere political principles, has always venerated the first written statement of those principles.

When I took my children to see one of the four original copies of Magna Carta in Lincoln Cathedral, there was no particular fuss, no line. But when the same copy was exhibited in New York in 1939, 14 million people crushed in to see it. The Second World War broke out while the parchment was still in the United States, and it was held in Fort Knox for safekeeping until 1945, the most fitting symbol imaginable of what the English-speaking nations were fighting for.

America was fortunate in the timing of its foundation. The first colonies were planted at the moment when the mania for Magna Carta in England was at its height. During the first three decades of the seventeenth century, while the pioneers were establishing themselves in Virginia and New England, English MPs and lawyers were pressing their claims against the Stuart kings, whom they regarded as having the same authoritarian tendencies that had roused the nation against King John.

Led by the brilliant jurist and MP Sir Edward Coke (pronounced "Cook"), they argued that the new dynasty had unbalanced the "ancient constitution": James I and, even more, his son Charles I were accused of raising taxes arbitrarily, sidelining Parliament, and violating the rule of law. In making these arguments, supporters of parliamentary prerogative explicitly linked their cause to that of previous campaigners against royal power. They ransacked libraries in search of medieval documents that might support their interpretation of the "ancient constitution." Again and again, they came back to Magna Carta.

The Englishmen who crossed the Atlantic during those turbulent years were unusually attuned to Coke's arguments. They
were drawn from the communities where radical sympathies
were strongest. Support for parliamentary supremacy was especially pronounced among Puritans. It also tended to be concentrated geographically in the southeastern and eastern counties of
England (Coke himself was from Norfolk, the large, sparse, flat,
Protestant county on the country's eastern edge). The first Americans, especially those who settled New England, came from precisely these groups.

They arrived in the New World keenly aware of their rights as
freeborn Englishmen, and drew consciously on the language of the
Great Charter when framing the charters of their colonies, notably
the Massachusetts Body of Liberties. In 1638, Maryland sought permission to recognize Magna Carta as part of the law of the province.
The earliest Virginia Charter, issued in 1606, was chiefly drafted by
Coke himself.

The first copy of Magna Carta to be printed on American soil
was published as early as 1687, as part of William Penn's *The Excellent Privilege of Liberty and Property: being the birth-right of the
Free-Born Subjects of England*. The founder of Pennsylvania was in
no doubt that Magna Carta set the English-speaking peoples apart
from other nations:

> In other nations, the mere will of the Prince is Law, his word takes
> off any man's head, imposeth taxes, or seizes any man's estate,
> when, how and as often as he lists. In England, each man has a
> fixed Fundamental Right born with him, as to freedom of his
> person and property in his estate, which he cannot be deprived
> of, but either by his consent, or some crime, for which the law has
> imposed such a penalty or forfeiture.

In time, there was to be a slight divergence between the British and American understandings of the Great Charter. In Britain, constitutional theory began to tend toward parliamentary sovereignty, and significance was attached to the fact that Magna Carta had given rise to representative government. The colonists, by contrast, clung to the belief that the Great Charter, as a written compact, stood above both Crown and Parliament—a belief that, in time, gave rise to the U.S. Constitution.

To be clear, American Patriots didn't just propose ideas that were inspired by the philosophy of Magna Carta. They saw that document itself as a part of their inheritance. When, as they perceived it, George III violated their patrimony, they took up arms to defend it.

The pamphleteers who made the "no taxation without representation" argument didn't claim that this was some natural or universal principle. Rather, it was an inherited part of the ancient constitution carried across the Atlantic by their ancestors, and could be found in Article 12 of Magna Carta: "No scutage or aid is to be levied in our realm except by the common counsel of our realm."

The first Continental Congress in 1774 explicitly linked its actions to those of the barons at Runnymede and the parliamentarians on the battlefields of the English Civil War. Its final Declaration and Resolves listed many of the same grievances addressed by Magna Carta: "The respective colonies are entitled to the common law of England, and more especially to the great and inestimable privilege of being tried by their peers of the vicinage [neighborhood], according to the course of that law." In breaching this and other inherited rights, King George's ministry was violating "the English constitution." The delegates were driven to resist their sovereign "as Englishmen their ancestors in like cases have usually done, for asserting and vindicating their rights and liberties."

In 1775, Massachusetts adopted, as its state seal, a Patriot carrying Magna Carta in one hand and a sword in the other.

After independence, the language and philosophy of the Great Charter found their way into the U.S. Constitution and, in particular, the Bill of Rights. The Fifth Amendment echoes the language of Magna Carta's Clause 29: "No person shall be deprived of life, liberty, or property without due process of law."

Magna Carta might not be formally on the law books in the United States, but it has been cited more than a hundred times by the Supreme Court.

In 1937, alarmed by what he saw as President Franklin Roosevelt's unbalancing of the U.S. Constitution, the great Texan jurist Hatton Sumners delivered one of the truest and pithiest statements of American political theory:

> There is a straight road which runs from Runnymede to Philadelphia. We did not "borrow" provisions from the British Constitution, which had come from the people; those provisions were ours, paid for with the lives of our ancestors on many a battlefield. I have examined the matter. I tell you our Constitution came up from the body of a self-governing people. But we can lose our capacity to govern by its nonexercise.

It used to be fashionable to claim that Magna Carta was "revived" in the seventeenth century, that it was conscripted to serve in a wholly unrelated constitutional dispute of that era, and that we shouldn't see it as anything more than a deal between a cornered king and his mutinous nobles—a deal that was promptly broken the moment the King could get away with it. Plenty of historians still take this line, terrified of saying anything that might seem to endorse a self-congratulatory, anachronistic, or Whig view of history.

Yet the stubborn fact remains that Magna Carta was quoted throughout the Middle Ages in precisely the way that Whig historians were later to claim: as a defense against arbitrary government. It was seized on by the tax-weary subjects of Edward I, who forced that martial monarch to reissue it in 1297. It was cited repeatedly during the fourteenth century in the cause of baronial or parliamentary supervision of the government. A statute of 1369, during the reign of the equally martial Edward III, declared Magna Carta to have constitutional force, overriding lesser laws: "If any Statute be made to the contrary, that shall be holden for none."

By the fifteenth century, Magna Carta had been reconfirmed by various monarchs more than forty times. The idea that Sir Edward Coke found a copy in some old collection, and gave contemporary relevance to a text that had until then been of antiquarian interest, depends upon disregarding a great deal of what was said during the intervening four centuries. It depends, too, on disregarding the most immediate practical consequence of Magna Carta, namely the establishment of an elected assembly whose duty was to hold the monarchy to its side of the bargain.

THE RESTORATION OF PARLIAMENT

A charter without enforcement mechanisms is worthless. This point might seem obvious, but it can be remarkably hard to get across. Our own age has seen a proliferation of human rights conventions and declarations, some national and some transnational, most of which serve, in practice, to shift power from elected representatives to unaccountable bureaucrats and judges, and thus to leave us less free. Yet I've never found a way to make this argument without looking as though I'm somehow lukewarm about the principle of human rights.

The precepts set out in, say, the European Convention on Human Rights, or the EU's Charter of Fundamental Rights and Freedoms, are not very different from those guaranteed by the constitutions of East Germany or the Soviet Union: freedom of speech, freedom of assembly, and so on. But, as the citizens of the communist states knew, paper rights have no value without instruments of democratic control. In the Anglosphere, rights were traditionally bound up with representative institutions—a connection that has its genesis in Magna Carta and the period that came immediately afterward.

The Great Charter had barely been sealed when King John disregarded it, plunging the country into a civil war. But, just as that war seemed to be on the point of stalemate, the providentially bad monarch rendered one last service to England by dying opportunely in Newark Castle in October 1216 (almost certainly of dysentery, and sadly not, as one source claims, from a surfeit of peaches).

His nine-year-old son was crowned King Henry III. Happily for England, the new king's mother had no interest in ruling through him during his minority, and power passed to a council of barons. We have records of assemblies—generally called parliaments from the 1230s onward—levying taxes, approving laws, debating foreign policy, and proposing candidates for public office. Here, in short, are parliaments that, in their functions if not yet in their composition, resemble those that have met ever since.

How unusual were these parliaments by contemporary standards? It's true that there were estates general and great councils meeting around Europe at the time, some of them under license of royal charters. But what happened in England was unique. England's Parliament was not a creature of monarchical will, but an arbitrator and guarantor of something that was above the monarchy, namely the law. It was peculiar in its legal status, in its representative character, and in its authority.

In 1227, a French letter writer contrasted the powers of the

French and English rulers. If the king of France wanted to make war, he wrote, he had to consult only two men: his chief adviser and his grand chamberlain. But if the king of England wanted to make war, he had to take counsel with many men in a prearranged assembly. It was an early version of what Sir John Fortescue, writing in around 1470, was to identify as the difference between England's "public and royal government" (*dominium politicum et regale*) and France's narrowly "royal government" (*dominium regale*).

The period of Henry III's minority was critical, because it established concrete precedent for the conciliar government foreseen by Magna Carta. Had John lived longer, the ideals of the Great Charter might not have been realized. Humble kings are rare creatures in the Middle Ages: few men can wield supreme power without being turned by it. A regency represented a unique opportunity to disperse power and, flushed with their success at Runnymede, the English barons seized it.

Sure enough, once he came of age, Henry III, a vacillating and self-centered man, chafed against the constraints of Parliament. He repeatedly violated both the letter and the spirit of the Great Charter, interfering with the common law, disregarding the council, exacting money arbitrarily from his subjects, and, as his father had done, filling his court with southern Frenchmen and other "aliens."

England's political classes, however, by now had firsthand experience of an alternative. The concept of consensual and collegiate government was not simply a promise on parchment; it was a recent memory. Opponents of monarchical absolutism had a very clear idea of what they wanted, namely formal power over some of the King's functions. And they had a mechanism to secure that objective: control of the national purse strings.

The confrontations between King and Parliament, in the thirteenth century as later, were chiefly financial. Henry III was constantly hungry for cash. He tried on more than one occasion to regain his father's lost French lands. He spent lavishly on a shrine to Edward the Confessor, whom he regarded as the founder of his line. He wanted to buy the throne of Sicily for his second son, Edmund. And he was naturally extravagant, both in his personal lifestyle and in what he regarded as the generosity proper to a monarch. Among other things, he gave London its first zoo, with an elephant, a rhinoceros, a polar bear that swam in the Thames, and lions (visitors were asked to bring a cat or dog for the lions as their entrance fee).

The sums involved in these schemes were, by the standards of the time, fantastic: £80,000 on an attempt to reconquer Poitou in 1242, £36,000 on an expedition to Gascony in 1253, £45,000 on Edward the Confessor's mausoleum and the rebuilding of Westminster Abbey around it, £90,000 on underwriting the pope's debts in exchange for pontifical backing for Edmund's Sicilian candidature. Parliament naturally resented authorizing such grants. When it did approve them, it usually held out for something in return. In 1225, for example, a great council approved a tax of one-fifteenth on movable goods to pay for a French campaign in exchange for the King reissuing Magna Carta. This pattern was to continue for centuries: the executive demanding a subsidy, the legislature debating the issue (for more than a week on this occasion) and eventually agreeing to grant supply in return for the redress of grievances.

These grievances multiplied during Henry III's reign, as he scorned both Parliament and the Great Charter. There were arbitrary expropriations "by the King's will"; royal charters that had been granted in perpetuity were revoked; heirs were excluded from their estates; heiresses were married to royal favorites; fines and exactions were used to raise revenue without due authority.

Supporters of Parliament responded by seeking, and eventually achieving, official status for their institution. A permanent baronial council was established, with a formal voice in the conduct of foreign affairs, royal appointments, and policy in general. At the same time, parliamentary meetings began to happen regularly, and to be held increasingly in one place: Westminster.

The king at first went along with these changes, greedy for the direct taxation that Parliament granted him in return. But in 1261, backed by the pope, he sought to reassert his power, prompting another civil war. The parliamentary reformers were led by the king's brother-in-law, a Frenchman called Simon de Montfort, whom later generations of British historians enthusiastically claimed as an English patriot, and whose image adorns the U.S. House of Representatives. Having defeated the Royalist armies, and captured the Prince of Wales (the future Edward I), de Montfort summoned a parliament to meet at Westminster Hall in 1265.

Historians have traditionally treated that assembly as the first session of the parliament that, to this day, meets on the same spot. What distinguished de Montfort's parliament from previous assemblies was that he invited every shire to send two "prudent and law-worthy" knights, and major towns to send two burgesses. Extraordinarily—and uniquely in Europe at that time—he asked that these representatives be directly elected, and as far as we can tell, the vote was open to every freeholder: it wasn't until two hundred years later that a tax threshold was imposed, and the franchise thereby restricted.

History rarely offers clear-cut beginnings and endings. Knights had certainly sat in some previous parliaments, and townsmen might occasionally have done so, too. Nor did Parliament meet uninterruptedly from that moment. De Montfort went on to lose the civil war, dying on the battlefield. The knights and townsmen did not meet again until 1275, and their sessions cannot be considered remotely regular until 1295. Not until the 1320s did the knights and burgesses begin to sit separately from the lords and bishops, and only following a judicial decision of 1489 was it established that laws needed to be passed by both houses, and not simply the upper chamber.

Even then, the status of the House of Commons (as we can by now call it) was far from secure. There were lengthy periods when no MPs sat at all, even as late as the seventeenth century: Charles I ruled without Parliament between 1629 and 1640, and would have dispensed with the institution altogether had he not needed revenue to finance his war in Scotland. The supremacy of the House of Commons over the House of Lords was established in law only in 1911, though it had been accepted in practice long before then.

Nor should we imagine that Parliament became steadily more representative. A combination of demographic changes and deliberate gerrymandering meant that the franchise could dwindle as well as increase. The House of Commons on the eve of the 1832 Reform Act was, on some definitions, a less representative body than it had been five centuries before. There were rotten boroughs, where depopulation had left just a few dozen residents with the vote. There were pocket boroughs, where a local grandee or, more often, the government of the day, effectively got to nominate the MP. The new industrial towns were, for all intents and purposes, unrepresented. Not until 1918 did all adult men get the vote; and not until 1928 was it extended to all adult women.

Nonetheless, Simon de Montfort can justly be considered the father of Parliament. If the House of Lords was born at Runnymede, the House of Commons was born half a century later when he called the elected knights and burgesses to Westminster. The pattern of counties and boroughs being separately represented continued until well into the nineteenth century. And the essential functions of the House of Commons—approving officeholders, debating policy, and, above all, controlling the purse strings—have continued more or less uninterrupted for nearly eight centuries.

"Parliament," wrote the intellectually dazzling Conservative MP Enoch Powell, "is a word of magic and power in this country." England's Parliament is not the world's oldest. Iceland's Althingi was established in 930, although its existence was not unbroken. The Tynwald on the Isle of Man has met uninterruptedly since 979.

What is peculiar to the English Parliament, which later became the Parliament of Great Britain and, later still, that of the United Kingdom, is the legitimacy and authority it was accorded. The country's great constitutional upheavals—the wars of John and Henry III, the depositions of Edward II and Richard II, the Reformation, the English Civil War, the Glorious Revolution, the Acts of Union—were experienced as parliamentary events.

At the beginning of the 1960s, Powell sought to articulate the essence of Englishness. He was wrestling with the loss of an empire of which, until then, he had been a keen supporter. Like other Britons of his generation, he had defined himself, at least in part, as an imperial citizen. As the colonies became independent, he craved a more elemental form of national identity, and searched for it in the years before the English-speaking peoples had crossed the oceans. Yet, when he found the essence of Englishness, it turned out to be precisely the thing that English-speakers had taken with them across the seas, and that was now the property of the Anglosphere as a whole: parliamentary government.

On St. George's Day, 1961, Powell gave a speech that imagined the English, their wanderings over, returning to their homeland to find the secret of their nationhood. The speech is worth quoting at some length, partly for its lyrical quality, but mainly because it expresses the unique centrality of Parliament in England's—later the United Kingdom's, later the Anglosphere's—story.

> Backward travels our gaze, beyond the grenadiers and the philosophers of the eighteenth century, beyond the pikemen and the preachers of the seventeenth, back through the brash adventurous days of the first Elizabeth and the hard materialism of the Tudors and there at last we find them, or seem to find them, in many

a village church, beneath the tall tracery of a perpendicular East window and the coffered ceiling of the chantry chapel.

From brass and stone, from line and effigy, their eyes look out at us, and we gaze into them, as if we would win some answer from their silence. "Tell us what it is that binds us together; show us the clue that leads through a thousand years; whisper to us the secret of this charmed life of England, that we in our time may know how to hold it fast."

What would they say?

They would speak to us in our own English tongue, the tongue made for telling truth in, tuned already to songs that haunt the hearer like the sadness of spring. They would tell us of that marvelous land, so sweetly mixed of opposites in climate that all the seasons of the year appear there in their greatest perfection; of the fields amid which they built their halls, their cottages, their churches, and where the same blackthorn showered its petals upon them as upon us; they would tell us, surely of the rivers, of the hills and of the island coasts of England.

One thing above all they assuredly would not forget; Lancastrian or Yorkist, squire or lord, priest or layman; they would point to the kingship of England, and its emblems everywhere visible.

They would tell us too of a palace near the great city which the Romans built at a ford of the River Thames, to which men resorted out of all England to speak on behalf of their fellows, a thing called "Parliament"; and from that hall went out their fellows with fur trimmed gowns and strange caps on their heads, to judge the same judgments, and dispense the same justice, to all the people of England.

Parliament was more than a place where representatives met to determine affairs of state. It was the supreme guarantor of the rule of law, the last defense of personal liberty, the chief exemplar of national exceptionalism.

That exceptionalism was present from the beginning. The first English-speakers had carried with them, from deep in the German woods, the idea that the law was the property of the tribe as a whole, and not an expression of the ruler's will. They brought, too, a tradition of open-air meetings at which the affairs of the tribe were debated and decided. They even had some notion, if Tacitus is to be believed, of what their descendants would call government by consent.

After the fall of Rome, we can see these ideas given expression in various Germanic kingdoms. There were councils and assemblies throughout Europe, their function as much judicial as legislative, in the sense that they adjudicated disputes as well as approving law codes.

Across the West, this tradition died out between the ninth and eleventh centuries. Kings and nobles did what rulers in every age and nation do: they used coercive force to rig the rules in favor of themselves and their descendants. The aristocracy became, legally, a separate caste, with unique privileges. The great landowners had almost arbitrary power on their estates. The kings were constrained, not by the law, but by the balance among their great magnates. Outside the towns, feudalism, serfdom, and inherited status were the norm.

Common law and popular assemblies survived only in a few parts of the Nordic world and, outstandingly, in England. What made England exceptional was its early development as a nation-state. This development, in turn, owed a great deal to geography. England was not exactly an island but, by the tenth century, it was overwhelmingly the preponderant state on the island of Great Britain, and Welsh and Scottish princes periodically did homage at the English court. Clear borders, a united national church, a single language, and a sense of common identity created what we would now call civic society. The government of the time was exceptionally powerful and wealthy, yet maintained a relatively small military force. It is no coincidence that the only other places with comparably antique parliaments are also islands: Iceland and the Isle of Man.

By the eleventh century, England was precociously enjoying constitutional government: its kings were required to maintain the law, as determined by the Witan. The justice system applied evenly to the greatest ealdorman and the meanest ceorl.

There was a major reverse after 1066, when the country was subjected to European feudalism. But, although landowning patterns, church government, parliamentary process, and royal authority were transformed, the Anglo-Saxon administrative system survived at local level. Shire and hundredal courts continued to operate according to a common law system that, in the mid-twelfth century, was elevated to national level. These courts kept alive the notion of representative government—or, at least, of decisions being taken consensually at public meetings.

Eventually, this notion infected the Anglo-Norman barons who, by the beginning of the thirteenth century, had come to think of themselves as Englishmen and to take an interest in the way that Englishmen had traditionally held their rulers to account. The result was Magna Carta and, as just as important, a national parliament to ensure that its terms were kept.

The road from those first parliaments in the thirteenth century to modern democracy was, as we shall see, often bumpy; at times it was little more than a dirt track. But let us leave that story for now, and turn to an aspect of Anglosphere exceptionalism that we have not yet considered, namely the peculiarities of social organization that distinguished English-speaking society from virtually the entire Eurasian landmass. It is here, as much as anywhere, that we shall find the origin of modern capitalism.

CHAPTER 4

Liberty and Property

Every Free-born Subject of England is heir by Birth-right unto that unparalleled privilege of Liberty and Property, beyond all the Nations in the world beside; and it is to be wished that all men did rightly understand their own happiness therein.

—WILLIAM PENN, 1687

The majority of ordinary people in England from at least the thirteenth century were rampant individualists, highly mobile both geographically and socially, economically rational, market-oriented and acquisitive, ego-centered in kinship and social life. Perhaps this is no surprise, for it makes them very like their descendants.

—ALAN MACFARLANE, 1978

ANGLOSPHERE EXCEPTIONALISM

In 2010, when I was briefly and uncomfortably a front-bench spokesman for my party in the European Parliament, I had to deal with a draft law to harmonize inheritance laws across the EU. The proposal was a technical one, to do with what happened to the

belongings of people who died intestate while living in another EU
state. But it revealed a striking difference between British and Continental notions of property rights.

In most of Europe, an individual cannot bequeath his possessions
as he pleases. There are laws that, for example, guarantee a portion
of his estate to surviving widows and children. In some states, you
have discretion over only a third of your assets at death.

In Britain, by contrast, as in most of the Anglosphere, you can do
more or less what you like with your property—at least, with what
is left over once death duties have been paid. If you want to leave everything to a trust to look after your cat, it's up to you. If you want to
alter your will in favor of your new teenage boyfriend or girlfriend,
that's your descendants' tough luck.

Wills and inheritances loom commensurately large in the Anglosphere consciousness. An immense number of English-language
novels turn on them. We have come to take the idea of total ownership for granted.

Yet our common-law attitude to property is, by global standards,
extraordinary, and requires some explanation. It is the approach
taken in most Roman law states that is, on the surface, more rational. Why should the wishes of a dead person override the needs
of the survivors? The living, after all, are better placed to see how
assets might usefully be allocated.

The answer goes to the heart of how different societies conceive
ownership rights. If all property is ultimately the tribe's or the kin
group's or the king's (or, nowadays, the state's), then it makes sense for
surviving members of the tribe (or, nowadays, government officials) to
determine the most rational distribution of resources. Ownership in
such societies is, in effect, a form of lease: an exclusive right to enjoy a
given asset. Such a right will not usually be prolonged beyond death.

Man is a social animal, and most human societies have conceived property as at least partly communal. Before there were laws,

before there were cities, before there were farms, before there were tools, even before there were words, men and women existed in kin groups. From prehistoric times until the early modern period (and, indeed, until the present day in large parts of the world), the basic economic unit was the extended family. The laws that governed ownership and transaction were developed in this context, and naturally elevated the clan over the individual.

The idea that property represented a total right of ownership by a single person was peculiar to English-speaking societies. The discrepancy I encountered in the European Parliament didn't simply set the United Kingdom apart from the rest of the EU; it set the Anglosphere apart from virtually the entire planet. Hence the misunderstandings and conflicts that accompanied so many colonial encounters between English-speakers and indigenous peoples.

When settlers accustomed to common-law property rights arrived in North America, in Africa, in New Zealand—even, many centuries earlier, in Ireland—they found that their notions of property were literally beyond the comprehension of local people. The settlers would, as they saw it, purchase land. The natives, however, had no concept of permanently alienating land from their tribe. In their eyes, they were selling the right to make use of resources on territory which an individual might no more own than he could own the wind or the sunshine.

We hear the same grievances over and over again in the complaints of the indigenous peoples, whether Cherokee, Maori, or Kikuyu. The settlers, they said, had cheated them out of their inheritance, or tricked them into land deals they hadn't properly understood, or bribed an individual chief into signing away rights that were not his to give. Some of the colonialists were, of course, unscrupulous and rapacious, and there were plenty of acts of straightforward theft. But there was also a fundamental incompatibility of understanding that made bloodshed almost inevitable.

For a long time, the unique nature of Anglosphere land law went unremarked. Consciously or otherwise, historians tended to see social history in broadly Marxist terms. According to Marxist orthodoxy, a peasant society had, at some stage in the past five hundred years, given way to a capitalist society. The former had been settled, rural, hierarchical, and largely autarkic; the latter was individualistic, specialized, money-based, and competitive.

Karl Marx himself was keenly interested in England as a template for his theories. It was, he believed, the country where capitalism had first developed, and so would be the first to succumb to a socialist revolution.

England, Marx taught, had been a peasant society until the end of the fifteenth century, when it began its gradual shift toward a market economy. By "peasant society," he meant that England had had an economy based on self-sufficient families and feudal dues rather than money. "The mode of production itself had as yet no specific capitalist character," he wrote. Land was treated as a common family birthright, not a commodity to be bought and sold. The basic economic unit was not the individual, but the peasant farm: "The economic totality is, at bottom, contained in each individual household, which forms an independent center of production for itself: manufactures purely as domestic secondary tasks for women etc."

The transformation, for Marx, came when rent-in-kind began to give way to money rents, something that led to "the formation of a class of propertyless day-laborers," who could then be exploited by employers.

Marx presented his economic models as empirical, scientific facts, not as political opinions. His followers duly followed his version of history almost to the letter. The easygoing Friedrich Engels wrote that England's "natural economy" in which "the exchanging family heads remain working peasants who produce almost all they require with the aid of their families on their own farmsteads," had

been displaced by a "capitalist economy" that had "dissolved all inherited and traditional relations and replaced time-hallowed custom
and historical right by purchase and sale." A hundred years later, the
respected leftist British historian Christopher Hill made precisely
the same claim, though he brought the date of the transformation
forward a little:

> In 1530, the majority of English men and women lived in rural
> households (mostly mud huts) which were economically self-
> sufficient: they wore leather clothes and ate black bread from
> wooden trenchers: they used no forks or pocket-handkerchiefs.
> By 1780 England was being transformed by the factory system:
> brick houses, cotton clothes, white bread, plates and cutlery were
> becoming accessible even to the lower classes.

What's striking, reading these words, is that they don't seem especially Marxist. Like many of Marx's precepts, this one has become
so orthodox in academic circles that we have lost sight of who first
authored it. Virtually every major writer followed Marx in positing a transformation from familial self-sufficiency to a modern
cash economy, though they differed a little as to precisely when the
change had come and what had caused it. Max Weber, the German
philosopher who has as good a claim as anyone to be the founder of
sociology, thought that Puritanism, which made a virtue of thrift
rather than of generosity, had been as important as the development
of a class of free laborers.

R. H. Tawney, the enormously influential British historian,
agreed with Weber, positing a shift from a "distributive" Catholic
ethic to an "acquisitive" Protestant one. Yet both writers, in common
with virtually every social historian who followed, unhesitatingly
accepted the Marxist account of a progression from self-sufficient
family units, in which all worked and shared, to a society of atomized

individuals mediating their relationships through cash. Tawney described the peasant household unit rather more eloquently than the Marxists whom he was echoing: "It is a miniature co-operative society, housed under one roof, dependent upon one industry, and including not only man wife and children but servants and laborers, ploughmen and threshers, cowherds and milkmaids, who live together, work together and play together."

It comes as quite a jolt to realize that this narrative is orthodox Marxism. Most of us, if we got any social history at school, will have been taught a version of it. It is reinforced in countless historical novels and dramas. It was played out before a massive global audience in the opening ceremony of the 2012 London Olympics, which showed a bucolic landscape, complete with happy shepherdesses, being enclosed by private owners, who then sent the peasants to work in their grim mills and chimney-stacked factories.

The reason that the Marxist account caught on, and spread well beyond the left, is that it was largely true—true, in fact, in every respect except one. It described a process through which almost every European country had passed, albeit some much earlier than others. It applied, not just to Europe, but to Russia, China, and India. Indeed, it was relevant to almost all European and Asian societies. The only place to which it didn't apply at all was the place Marx had been writing about: England.

WHERE WERE ENGLAND'S PEASANTS?

We have already touched on the Peasants' Revolt of 1381. There are various ways to think about that bloody episode: as a protest against the conditions of serfdom; as a consequence of the rising wealth and rising aspirations of the class of yeomen whose influence had grown since the depopulation of the Black Death; or as the last

Anglo-Saxon revolt against a French-speaking aristocracy. The one thing that absolutely no one called it at the time was a peasants' revolt, for the good reason that the word *peasant* didn't exist in English. Or, to be precise, it was used only when talking about foreigners, being a direct rendition of the French word *paysan*, countryman.

The reason the word didn't exist in English is that the thing it described didn't exist in England. The word *peasant* was understood by contemporaries much as it is by historians. It didn't simply mean someone who dwelt in the countryside. It also denoted the socioeconomic characteristics identified by Marx, Weber, Tawney, and the rest.

A peasant was attached to, and to an extent defined by, his family's land. That land was not his to do with as he wanted: it was the common inheritance of his extended household. It could be sold only in times of dire need, and with the consent of all male heirs. The peasant's family unit tended to produce and consume internally, with little trade. Staples would be grown or reared on the land, and only luxuries and occasional goods exchanged. Even then, the exchange was often in the form of barter rather than cash purchase. Money existed, but tended to be treated as one more asset to be stored, not as a means of exchange.

Such peasant societies existed medievally right across Europe and Asia. They began to adapt in northwestern Europe from the sixteenth century, but were still widespread in Eastern Europe until the nineteenth, and in Russia until the twentieth.

Until recently, it was assumed that England's rural economy had not differed substantially from those of neighboring states. Then, in the 1970s, a young Cambridge historian named Alan Macfarlane began to study the parish records of medieval England. What he found, to his astonishment, was a form of social organization that met none of the criteria of what is generally called a peasant society. The commonalities of rural society ran right across Eurasia, from the Pacific to the Atlantic. But they stopped at the English Channel.

In most of Europe, landownership was settled, with farms being treated as an inalienable patrimony. In England, by contrast, there was a lively land market from at least the thirteenth century (earlier records are harder to come by). In most of Europe, children would work on their parents' farms, receiving board and lodging rather than wages. In England—to the surprise and occasional disgust of overseas visitors—children would generally have left the family home by their teens, either for apprenticeships or to work elsewhere. The farmwork would instead be done by hired hands for competitive pay. In most of Europe, the family was recognized as the primary unit, not just in custom but in law: parents generally could not disinherit their children, and the family plot was treated as a communal resource. In England, there was almost no notion of shared ownership. A boy who had reached legal maturity was, in the eyes of the law, a wholly free agent: his father had neither claims over him nor duties to him.

Macfarlane's thesis was at odds with the temper of the times. In the 1970s, most academics were still influenced by Marxist historiography, for all that they might have consciously eschewed Marxist politics. But he found his conclusion to be inescapable: "What now seems clear is that England back to the thirteenth century was not based on either 'Community' or 'communities.' It appears to have been an open, mobile, market-oriented and highly centralized nation, different not merely in degree but in kind from the peasantries of Europe and Asia."

When and why did these differences emerge? On the question of when, Macfarlane frankly admitted that it was impossible to say. The peculiar individualism of English society was taken for granted in the earliest sources he could find. He suspected that the origins went right back to the earliest Anglo-Saxon settlements, and had their roots in the first-century Germanic commonwealths described by Tacitus. But, in the absence of hard evidence, he could only guess.

On the question of why, he was much clearer. The individualistic character of English society was guaranteed by two connected legal features: primogeniture and the total ownership of land by a single person.

Primogeniture—the practice of settling an entire estate upon the eldest son, rather than dividing it equally, or passing it on collectively— is intimately linked to the alienability of land. It amounts in one sense to the disinheriting of all claimants in favor of a single one. Neither primogeniture nor outright ownership is compatible with a peasant society on the Continental model. Both are products of the bottom-up common-law system. Together, at a physical as well as a political level, these two precepts shaped the Anglosphere.

LAWS MAKE LANDSCAPES

Suppose you were flying from, say, Budapest to Birmingham, England, and dozed off. One glance from the window when you woke up would tell you whether you had crossed the English Channel. The land laws of England have taken tangible form in the countryside. Farms across Europe tend to be rectilinear, often divided into strips. So as not to waste space, the boundaries are generally marked by wire fences, which may be swiftly moved when brothers inherit.

In England, by contrast, the fields are irregular and undulating. They follow natural features, curving with brooks, rather than forming straight lines. They tend to be enclosed by more permanent barriers: hedgerows in most of the country, drystone walls in parts of the north and west.

England's landscape was made by English laws. Because properties did not have to be shared or divided, natural boundaries remained in place.

A countryside marked by hedgerows is one in which property has long been secure. Unlike a wire fence, a hedgerow cannot easily be moved. An old English hedge is a thick, bristling riot of plant varieties: dwarf oak, maple, honeysuckle, clematis, wild rose, sallow, blackthorn, hazel, alder. There is even a handy rule for calculating a hedgerow's age: count the number of species in a thirty-yard stretch, not including ivy or blackberry, and multiply by 110.

Such ancient boundaries, some of which have stood since Saxon times, tell us that these were not lands divided among siblings. As fortunes rose and fell, families would buy and sell whole estates. Here was a functioning land market, underpinned by confidence of tenure.

The same is true of the great estates that are such a marked feature of the English countryside. It is extraordinary, walking about their grounds today, to think that their eighteenth-century owners laid out their seedlings in such a way as to reach perfection hundreds of years later. What assurance those early landscape gardeners had in the stability of the nation and its political system. They expected their parks to be enjoyed, in their ripeness, by their great-great-great-grandchildren. They felt certain that their homes wouldn't, in the meantime, have been expropriated by a tyrant, wrecked by a mob, or used to billet foreign invaders.

These great estates were more than houses and gardens. To generations of Englishmen, they represented everything that was healthy and free about their nation. Some of the greatest writers of the sixteenth and seventeenth centuries—Ben Jonson, Thomas Carew, Andrew Marvell—are known as the Country House Poets, because they used the rural garden as a political metaphor, an antidote to court politics. The royal court, in their verses, was effeminate, mannered, artificial, scheming; the country was artless, organic, loyal, frank. The great country houses were, as a later generation would see it, a solid expression of Whig political philosophy.

One of the finest examples is not in England at all, but in Virginia. It was named Mount Vernon, after a British admiral, and in its frugal stonework and modest vegetable patches we can feel the character of its great yet humble proprietor, George Washington. When David McCullough wrote that Washington had left no autobiography, but had left Mount Vernon instead, it was no figure of speech.

Washington was a man of action, not given to overanalysis, but the English Whig tradition was bred in his bones. He founded the American Republic with an act of supreme renunciation, perhaps the greatest of all time, turning down the powers that his countrymen pressed upon him because he believed that to leave the court and return to your farm was the greatest virtue in public life. Here was the ideal of the Country House Poets made flesh.

"As the sword was our last Resort for the preservation of our Liberties," Washington told Americans, "so it ought to be the first thing laid aside when those Liberties are firmly established." As so often, the English-speakers who had crossed the Atlantic had brought with them a stronger dose of exceptionalism than those who stayed behind and, in their new home, distilled it to yet greater potency.

The settlers had carried an absolute commitment to private property—and, though the phrase was not yet current, to the free enterprise system implied by private property. William Blackstone, whose massive influence in the colonies we have already noted, described "sacred and inviolable rights of private property" as an "absolute right, inherent in every Englishman." John Locke, the other intellectual godfather of the American Revolution, went so far as to assert that "the great and chief end therefore, of Men uniting into Commonwealths, and putting themselves under Government, is the Preservation of their Property."

Belief in the absolute sanctity of ownership, not just in the sense of being allowed to settle your goods on whom you pleased, but also in the sense of free contract and minimal taxation, was common to almost all the Founders. But, unsurprisingly, they had less time for the second defining feature of English property law, namely primogeniture.

The Americas had largely been peopled by younger siblings. The grander families of Virginia—including the Washingtons—were known as the "Second Sons." Many of the younger brothers who had founded their lines in the New World had borne with them a sense of injustice that they had been denied any share of their ancestral lands through an accident of timing. Like Edmund in *King Lear*, they felt that the primogeniture rules were at odds with natural justice:

> Thou, nature, art my goddess. To thy law
> My services are bound. Wherefore should I
> Stand in the plague of custom and permit
> The curiosity of nations to deprive me
> For that I am some twelve or fourteen moonshines
> Lag of a brother?

English primogeniture had been, for all but a handful of aristocratic families with entailed lands, a tradition, not a legal obligation. Fathers were perfectly at liberty to disinherit their eldest sons if they chose. But traditions—what Edmund calls "the plague of custom"—matter, and Americans were determined to extirpate this one. Thomas Jefferson, who often quoted *King Lear*, revised Virginia's law codes so as to eradicate the slightest possibility of primogeniture in order that, as he put it, "every fibre would be eradicated of antient or future aristocracy." He wrote with feeling. His mother's family, the Randolphs, had come to Virginia in the person of a textbook "Second Son": William Randolph, younger son of a Royalist Warwickshire gentleman.

The abandonment of primogeniture eventually spread from the United States to the rest of the Anglosphere. By the late twentieth century, the practice continued only in a small number of aristocratic families. In 2012, the United Kingdom and the other Anglosphere and Commonwealth states that shared a monarch agreed without any fuss to alter the succession law so as to remove the preference for male heirs, opening the door to a number of claims by aristocratic elder daughters. For most families, the idea that a firstborn male heir was entitled to a greater share than his siblings had long since been abandoned.

Yet, while it lasted, the practice of primogeniture had vast social consequences. In Britain, unlike in Europe, the nobility never became a closed class. Younger sons of landowners would have to make their way in the world as soldiers or doctors or clergymen or businessmen. In most of Europe, where nobility was an inherited legal status, the aristocracy could comprise a substantial proportion of the population: as much 30 percent in some countries. In Britain, their numbers were always tiny. On the eve of the 1789 revolution, there were 140,000 aristocrats in France. In Britain, prior to the creation of life peerages in the 1960s, there were generally fewer than two hundred members of the House of Lords, and never more than six hundred.

A consequence was to make Britain a country with unusually high social mobility. That phrase is used nowadays by politicians and commentators to mean that poverty should be no bar to success. Yet this, if you think about it, is only half the picture. In a society with high social mobility, poor children will indeed be able to rise higher than in one where status is determined by birth. Yet one individual's ascent of the social scale must be matched by another's decline. We are talking here, not of absolute wealth, which can rise for everyone, but of someone's status relative to his fellow citizens. If a yeoman farmer becomes an earl, then several people in between will have moved fractionally down the scale.

This downward social mobility was enormously exaggerated by the primogeniture rules. One son would get everything; the rest would be cut adrift. As they moved down the social scale, they would carry with them the habits that they had picked up in child-hood: literacy, for example.

There is an intriguing theory, due to the science writer Matt Ridley (coincidentally a British hereditary peer, and thus a bene-ficiary of the primogeniture system), that such downward social mobility was the proximate cause of Britain's takeoff in the eigh-teenth century. That takeoff was preceded by a quite extraordinary demographic alteration. Put simply, from the beginning of the sev-enteenth century, the rich began massively to outbreed the poor.

In 2004, two academics at the University of California under-took a major survey of English wills at the turn of the seventeenth century. The results were striking. A man leaving less than £10 in his will would, on average, be survived by two children; a man leav-ing more than £500 would, on average, be survived by four. These were cruel times: medicine was primitive, hunger not uncommon, and infant mortality high. But, as incomes rose, the wealthier classes were able, in effect, to purchase a higher survival rate for their chil-dren. Since only one of these children could inherit the family estate, the rest would have to make their way in the world.

The seventeenth century was a period of what the French call *déclassement*: most Englishmen and Englishwomen had lower status than their parents. Many who had benefited from a measure of ed-ucation had to make their way as small businessmen or artisans. In consequence, literacy rates among these groups—at least as mea-sured by the number of people able to sign their names on legal documents—began to soar. In 1600, 35 percent of Englishmen were able to read. By 1700, the figure was 60 percent (25 percent for En-glishwomen). The population was primed and ready for the massive economic changes that were just getting under way.

Although primogeniture has now almost entirely died out in Anglosphere societies, its legacy has not. English-speaking socie-ties, on every continent, retain the peculiar concept of inalienable property that had been an almost unique feature of England itself.

The idea that the rights of an individual linger even after death—the ultimate defiance of collectivism—had profound political consequences. It facilitated the establishment of trusts and foundations: bodies that were there, in effect, to give force to the wishes of deceased property owners.

These institutions, in turn, helped create what we now call civic society: that space between the state and the individual that is filled by unofficial, voluntary, and philanthropic endeavor. Continental visitors to the Anglosphere were constantly struck by the extent to which private foundations took on many of the responsibilities that, in their countries, were discharged by the state, or at least by state-run churches. Endowments led to the creation of schools and hospitals, art galleries and orphanages.

These institutions, in turn, helped create a political culture in which charitable and nonprofit activity was no more seen as the government's responsibility than commercial activity. Even today I encounter the difference in the European Parliament. In Britain, there is a presumption of legality; across much of the Continent, it is assumed that you need to go to the authorities for a license before launching any initiative. The first reaction of my fellow European Parliament members, on learning that some new activity has escaped the attention of the state, is often to propose a pan-European regulatory regime. The roots of statism claw their way deep into the cold soil of the Middle Ages.

WHAT'S SO SPECIAL ABOUT CAPITALISM?

Some within the Anglosphere have always preferred the Continental model. Only the state, they argue, can be relied on to provide an even and reliable service where there is no profit motive. The trouble with depending on private beneficence, they go on, is that it puts too much discretion into the hands of wealthy individuals. What if they miss something? What if their judgment is awry? What if they capriciously decide that the recipient is no longer deserving? As the Labour politician Clement Attlee argued in 1920, "charity is a cold gray loveless thing. If a rich man wants to help the poor, he should pay his taxes gladly, not dole out money at a whim."

In fact, Attlee had it precisely the wrong way around. It is hard to think of anything colder, grayer, or more loveless than the modern welfare state. As well as being tailored to the circumstances of the recipient, charity allows the donor to make a moral choice. There is virtue in deciding to give away your money, but none in having the same amount taken from you through the tax system.

The Anglosphere conception of liberty has had its critics, domestic and foreign, down the centuries. Yes, capitalism might make people wealthier, said these detractors, but was there not a price to pay? Did people not lose an element of their humanity? Did they not become more selfish, more atomized, more calculating?

No. There is nothing selfish about capitalism. Like every economic model, it is a matrix within which individual actors can behave morally or immorally. But it does have one exceptional virtue: no one has yet come up with a system that rewards decent behavior to the same extent.

In an open market based on property rights and free contract, you become wealthy by offering an honest service to others. I am typing these words on a machine developed by the late Steve Jobs. He gained from the exchange (adding fractionally to his net wealth) and so did I (adding to my convenience).

Under the various forms of corporatism tried elsewhere, someone else—generally a state official—gets to allocate the goodies, guaranteeing favoritism and corruption.

That's not to say, of course, that malpractice is unknown in capitalism. Man is fallen and, under any system, some will give in to temptation. It's simply that, in a state-run economy, corruption is systemic and semilegal. Indeed, the most egregious forms of wrongdoing in Anglosphere economies tend to be the ones that involve governments: lobbying for improper favors, securing taxpayer bailouts, and the like.

Greed—that is, the desire for material possessions—is not a product of markets, but a product of a human genome evolved in the competitive environment of Pleistocene Africa. Capitalism harnesses greed to socially productive ends. The way to become rich in a free economy is to give others what they want, not to suck up to those in power.

A common criticism of the Anglosphere's economic system is that it elevates efficiency over every human virtue: faith, kindness, loyalty, decency.

In fact, it is difficult to think of a more ethical relationship than one created by a free contract. Each party will add to the other's well-being by doing precisely what is expected of him. The same is rarely true between individuals in other contexts, however benign their intentions. Even close friends—even husbands and wives— will sometimes fall short of each other's expectations.

You might be thinking that what I've written seems rather mean and mechanical. Nothing wrong with contracts, you might say: paying your staff on time, delivering the goods your customer wants, all perfectly good stuff. But it hardly competes with the kind of behavior for which there is no material reward: working in a soup kitchen, visiting prisoners, sending money to famine-stricken territories—or, indeed, simply being a good parent, neighbor, or friend.

And, of course, you're absolutely right. But that's not the choice. We're not arguing about whether generosity and charity are laudable. We're arguing about whether big government encourages them.

In his 2008 book, *Liberal Fascism*, Jonah Goldberg touched on a number of studies suggesting that people who believe in small government tend to give a higher proportion of their income and their time to charity than those who believe in big government.

It makes sense. Once you've established your fundamental decency by signaling your disapproval of the right things, why go any further? Once you've called for tax rises, why give to charity?

It can't be repeated too often: when you give to good causes, you are making a moral choice; when the government takes an equivalent sum from you in taxation and spends it on your behalf, you are not.

This argument barely needed making in the heroic years when the Anglosphere was beginning its ascent to greatness. Four centuries ago, it was taken for granted that liberty, property, and private virtue were interconnected. The unique emphasis on ownership in England and North America was regarded as a bulwark against tyranny and an invitation to private benevolence.

English-speakers created propertied democracies in every territory they settled except those parts of the West Indies that were run as slave plantations. Again, it is instructive to compare North America with South America. In the former, almost everyone was given the opportunity to acquire land, including indentured laborers at the end of their terms. As the American Republic expanded, successive governments did their best to encourage private ownership through a series of bills culminating in the Homestead Act of 1862, designed to people the frontier with smallholders by making new land, in effect, free. By the beginning of the twentieth century, 75 percent of Americans in rural areas owned land. In Argentina, the figure was 25 percent, in Mexico 3 percent.

The English-speaking peoples entered the early modern period with all the tools that were to lift them to the global hegemony that lingers in our own day: common law, sanctity of contract, representative government, liberty of conscience, secure property, personal freedom. Educated English-speakers at the time were conscious that their way of ordering their affairs set them apart from other nations. Some were intensely proud of that distinctiveness; others felt drawn to European alternatives.

During the seventeenth and eighteenth centuries, these two tendencies were to harden into opposed factions. On more than one occasion, they would come to blows. In the 1640s and again in the 1770s, the two sides would be driven to settle their differences through bloody wars. Happily for the human race, those conflicts ended in victory for the side that cherished the exceptional nature of the Anglosphere's political inheritance. It is to that story that we now turn.

CHAPTER 5

The First Anglosphere Civil War

In all Christian kingdoms you know that parliaments were in use an-
ciently, until the monarchs began to know their own strength, and, seeing
the turbulent nature of their parliaments, at length they by little and little
began to stand upon their prerogatives, and at last overthrew the parliaments throughout Christendom, except only here with us.

—SIR DUDLEY CARLETON, ROYAL MINISTER, 1626

By natural birth all men are equally and alike born to like property, liberty and freedom; and so we are delivered of God by the hand of nature
into this world, every one with a natural, innate freedom and property.
Even so are we to live, everyone equally and alike to enjoy his birthright
and privilege.

—RICHARD OVERTON, LEVELER LEADER, 1646

THE FIRST LIBERTARIANS

Burford is a quiet, honey-colored Cotswolds town, especially nota-
ble for its parish church, a marvel of twelfth-century craftsmanship
that, unusually in England these days, is full almost every Sunday.

In 1649, that church was the scene of a grisly execution. The second installment of the English Civil War had ended in total defeat for the Royalists, but some of the parliamentary soldiers were not satisfied. They believed they had been fighting to restore England's ancient constitution: to "throw off the Norman Yoke," in the popular phrase of the time. They understood the ancient constitution to mean parliamentary sovereignty, based on something close to one man, one vote. Yet it was already clear that the new regime had an altogether less radical agenda.

The MPs, who had been sitting since 1640, had long since exhausted their mandates, but seemed in no rush to hold fresh elections. The leader of the parliamentary forces, Oliver Cromwell, had little time for the democratic utopianism of some of his troops. Already the Puritan general was beginning to see himself as a Davidic figure, raised by God to deliver his nation. A number of his veterans concluded that, for all the republican rhetoric of the war years, they had exchanged one oligarchy for another. They began to organize, holding mass meetings and publishing pamphlets demanding the completion of the democratic revolution.

Eventually, Cromwell decided to take on the agitators, who had become known as Levelers from their belief that all men were equal. Three hundred of the most troublesome troops were incarcerated in Burford Church. Some whiled away the time by carving their names into the lead of the baptismal font, plaintive graffiti that can still be seen today. (The font remains in regular use: I recently stood godfather there to a beautiful little girl.)

Threatened with execution for insubordination, most of the imprisoned Levelers disavowed their radical views. But three refused to recant, and were dragged into the churchyard and shot—an act whose cold-bloodedness disturbed even the morbid sensibilities of mid-seventeenth-century England.

The place where they fell before the muskets has now become a pilgrimage site for the radical left. Many British socialists admire the readiness of the Levelers to defy authority, whether princely or ecclesiastical. They cheer their egalitarianism and their support, in an age when such things were almost unthinkable, for a mass franchise.

Yet the Levelers were no socialists. On the contrary, they might aptly be considered proto-libertarians. The starting point of their philosophy was the freedom of the individual. As he owned his own mind and body, so he had a right to the fruit of his labor. The individual, Levelers argued, anticipating John Stuart Mill by more than two centuries, should be able to do anything that does not prejudice the freedom of another. As Richard Overton put it in his 1646 pamphlet, *An Arrow Against All Tyrants*:

> To every individual is given an individual property by nature not to be invaded or usurped by any. For every one, as he is himself, so he has a self-property, else could he not be himself. Mine and thine cannot be, except this be. No man has power over my rights and liberties, and I over no man's. I may enjoy myself and my property but presume no further; if I do so, I am an encroacher and an invader upon another man's right—to which I have no right.

Here is an extraordinary sentiment for its time, seeming to anticipate the doctrines of such twentieth-century libertarians as F. A. Hayek and Murray Rothbard. And, indeed, these philosophers gladly acknowledged the Levelers as their forerunners.

According to Hayek, the Levelers had played an important role in the development of what he called Anglo-Saxon liberalism, taking unwritten notions of liberty and property and proposing a written constitution and a formal separation of powers.

"The Levelers were the world's first self-consciously libertarian mass movement," wrote the apostle of small government and Austrian School economics, Murray Rothbard. "John Lilburne, Richard Overton and William Walwyn worked out a remarkably consistent libertarian doctrine, upholding the rights of self-ownership, private property, religious freedom for the individual and minimal government interference in society."

As well as looking forward to modern libertarianism, the Levelers looked back to the lost age of Anglo-Saxon freedoms, which they believed to have been extirpated by the Norman Conquest—a view that, while a touch romantic, was, as we have seen, not entirely unfounded.

Those patriotic soldiers are the golden link in the chain that connects our contemporary Anglosphere freedoms to their prehistoric beginnings. Indeed, the more we read of Leveler doctrines, the harder it is to understand why socialists ever tried to claim them.

Perhaps it is a straightforward case of mistaken identity. There *was* a contemporary group known as the Diggers, sometimes confusingly called "True Levelers," who espoused proto-socialist policies, including the common ownership of land. These are men whom the contemporary left might in good conscience venerate.

But the Levelers were Euro-skeptic, tax-cutting, anti-state patriots. They believed in democratic control of state appointments; an end to the patronage of the executive; freedom from overseas entanglements; an accountable judiciary; free trade; and the absolute sanctity of property. While they supported a massive extension of the franchise, they were also clear that the vote should not be given to benefits claimants—or, as they put it, to those "in receipt of alms." Theirs was the philosophy that was to lead, by way of John Locke, to the U.S. Constitution and the modern Anglosphere imperium.

Consider some of the demands set out in the Leveler manifesto, *An Agreement of the Free People of England*:

That it shall not be in their [MPs'] power to make any laws to abridge or hinder any person from trading or merchandising . . .

That it shall not be in their power to continue excise or customs upon any sort of food or any other goods, wares, or commodities, being both of them extreme burdensome and oppressive to trade . . .

We therefore agree and declare that it shall not be in the power of any representatives to level men's estates, destroy property, or make all things common.

The Levelers did not invent the idea of representative democracy, nor of personal freedom, nor yet of inviolable ownership. They were heavily influenced by Edward Coke's reverence for Magna Carta and, like many who are now called radicals, they were, in their own eyes, conservatives, seeking the restoration of what they took to be the ancient and natural English constitution. Yet they did something that no one had attempted before. They grabbed the inchoate traditions of English liberty and pressed them together into a single, coherent program of government.

Theirs were elevated and ennobling ideals—ideals worth dying for. Burford lies within both my European district and David Cameron's parliamentary constituency. The prime minister, an orthodox, if unusually gifted, conservative, takes the conventional view that the Levelers were leftist agitators. But, properly understood, they were the first conscious proponents of the individualist philosophy that raised the English-speaking peoples to eminence. I never pass Burford now without visiting the old church, pausing by the spot where the men fell before the executioners' matchlocks, and bowing my head for a few moments in memory of three Anglosphere heroes: Cornet Thompson, Corporal Perkins, and Private Church.

POWER TO THE PEOPLE

The Levelers were ideological outliers, but they were living at a time when mainstream political thought was moving vertiginously. Four months before the Burford shootings, the English people had beheaded their king.

Charles I was not the first English monarch to be put to death by his subjects: his predecessors Edward II and Richard II had both been quietly disposed of. But the execution of Charles I was an altogether different matter. The king was not secretly murdered. He was put on trial, found guilty of various abuses, and then, in an act that sent shock waves through Christendom, condemned in open court.

Who had the power to order such a thing? By what right might subjects raise their hands against their sovereign?

The unhappy monarch asked precisely that question at his trial: "I would know by what power I am called hither. I would know by what authority, I mean lawful; there are many unlawful authorities in the world; thieves and robbers by the high-ways. Remember, I am your King, your lawful King." One of the judges, John Bradshaw, replied that kings, too, must obey the law, and that Charles I stood accused of breaching the covenant between ruler and ruled: "There is a contract and a bargain made between the King and his people, and your oath is taken: and certainly, Sir, the bond is reciprocal."

This was an accurate enough summary of the doctrine of government by consent which, as we have seen, had pre-Norman origins. But it didn't answer Charles's question. By what authority did the judges who sat before him claim the right to determine whether he had been in breach of his royal duties? Who gave them the power to decide?

The question was an especially delicate one because the handful of MPs still left in 1649 could claim only the most tenuous legitimacy:

there had been no elections for nearly nine years, and the members thought least likely to approve the trial of the king had recently been purged by the army. In any case, few of those who had taken Parliament's side in the recent civil war had tried to argue that the House of Commons had ultimate sovereignty. Their position, rather, was that supreme power rested with "the King-in-Parliament," Parliament being understood to be the two ancient chambers.

On this occasion, though, the House of Lords—which had been reduced by desertions to no more than a dozen members—flatly refused to approve the ordinance providing for the king's trial. MPs therefore turned openly to the Leveler doctrine that sovereignty was vested in the people, and expressed through their elected representatives: "The Commons of England, in Parliament assembled, do declare that the People are, under God, the Original of all just Power: And do also declare, that the Commons of England, in Parliament assembled, being chosen by, and representing the People, have the Supreme Power in this Nation."

To us, this seems a remarkably forward-looking statement, anticipating the democratic theories of our own time. Yet, at the time, it was seen almost precisely the other way around.

The temptation, when reading history, is to assume that anything that represents a move toward our present values and institutions is progressive, and anything that represents a move the other way is backward. But, of course, the people we are studying have no notion of what the constitutional arrangements will be in our day. In 1649, as in 1941, democracy was seen as anything but a coming force. The progressive, radical, forward-looking idea was monarchical absolutism. All over Europe, representative assemblies—diets, councils, estates, and corteses—were being sidelined or scrapped altogether. Centralization was seen as a modern force, a tidying away of the various local particularisms that held countries back. The idea of a religious duty to a divinely

appointed prince was very much in vogue—upheld as much by Lutherans and Calvinists as by Catholics.

The words of Sir Dudley Carleton that open this chapter were
not a warning against the subordination of Parliament but, on the
contrary, a celebration of monarchical right. It was a matter of some
regret to this gentleman, who had traveled widely in Europe in the
service of his king, that England remained stuck with its medieval
parliamentary system.

Across the Continent, seventeenth-century rulers swept aside
what few restraints had been placed on their power, and put in their
place the novel doctrine of the divine right of kings. Peter the Great
in Russia, Frederick William in Prussia, Charles XI in Sweden, and,
above all, Louis XIV in France constructed the elaborate machinery
of autocratic rule, complete with fiscal independence and juridical
supremacy.

In 1614, while English MPs were attacking the excesses of royal
spending in the most hectoring and ill-tempered manner, the
French Estates-General, which had never been more than a weak
advisory body, were being dissolved, not to meet again until 1789.
In 1653, when England was at the height of its republican interlude,
the Diet of Brandenburg met for the last time and formally surrendered what was left of its tax-raising powers to the monarch. In
1665, when Charles II was discovering that he was as financially dependent on Parliament as his father and grandfather, Denmark adopted the "King's Law," authorizing the sovereign to close down any
alternative centers of power and declaring that "he shall from this
day forth be revered and considered the most perfect and supreme
person on the Earth by all his subjects, standing above all human
laws and having no judge above his person, neither in spiritual nor
temporal matters, except God alone."

England and Scotland, almost uniquely, spent the seventeenth
century traveling in the opposite direction. In the furnace of their

shared resistance to the Stuart dynasty, whose heat was also felt in Northern Ireland and in New England, was forged the Anglosphere in its present political form.

The accession of James VI of Scotland as James I of England in 1603 marked the moment when the English-speaking peoples came together in a single regime, though a formal union between the two kingdoms would not follow for another century. Nonetheless, even combined, James I's realms weighed little in the scales of Europe, and showed no sign of defying the general trend toward monarchical absolutism. By the time his grandson James II was deposed in 1688, everything had changed, and the Anglosphere was on the road toward wealth, freedom, and world dominance.

In 1689, Britain received the closest thing it has to a written constitution, a charter that much resembled its American successor, a document known today as the Bill of Rights. Unlike Magna Carta, the Bill of Rights was consciously conceived as a constitutional settlement. And, unlike Magna Carta, it provided for a form of parliamentary sovereignty that went well beyond baronial councils. It is therefore worth taking a moment to reprise the story of the anti-Stuart struggle that brought it about, which united the English-speaking peoples for the first time and left them with the form of parliamentary government they have known ever since.

SECTARIANISM, SUBSIDIES, SOVEREIGNTY

Throughout the seventeenth century, Englishmen tended to look back with fond nostalgia to the reign of Elizabeth I. Good Queen Bess was presented as having been everything her Stuart successors were not: properly Protestant, exuberantly patriotic, devoted to her duty.

Such wistfulness tends to be stronger in retrospect. There were few expressions of genuine sorrow when the old virgin queen, who had acquired a slightly pantomime appearance with her wigs and face powder, died in 1603, to be succeeded by her cousin, James VI of Scotland. Then the mood was upbeat: a tough-minded and likable king, with achievements already to his name, would, Englishmen said, bring to an end the half century of "petticoat government" that had begun with Mary I's reign in 1553.

James was unlucky in the timing of his accession. In 1598, France had granted toleration to its Protestant subjects, which lessened Englishmen's sense of being under siege and paved the way for an end to the Anglo-Spanish war in 1604. That war had brought a sense of unity and common purpose to Elizabethan England, as well as an outlet for the energies of its pugnacious people. Those energies were now turned inward.

In the aftermath of the peace settlement, James, who had a lifelong fascination with theology, hoped to lift the penal laws against English Catholics—laws that were, in any event, being halfheartedly and infrequently applied. Indeed, he dreamed of a grand reconciliation whereby the pope would recognize his control over the English church in return for his acknowledgment of the pope as "primus episcopus inter omnes episcopos."

Such a deal was acceptable neither to the Vatican nor to English Puritans. James was forced to drop his plans for toleration, and the sense of dashed expectations tempted a handful of English Catholics into the madness of the Gunpowder Plot, already discussed. The foiling of that terrorist outrage brought the monarch some short-lived sympathy, but it was not long before the quarrels between King and Parliament resumed in earnest.

These quarrels turned on three great issues: money, religion, and power. Elizabeth had had to sell off a portion of the Crown lands, reducing the monarchy's regular income and making her successors dependent on parliamentary subsidies. There was also a rise in inflation throughout Europe in the seventeenth century, which reduced the value of the Crown's fixed rents. Unfortunately for the King, his fiscal reliance on MPs was increasing at the very moment when the end of the Spanish war had exhausted their generosity.

Arguments about money are generally ill-tempered, and arguments about religion rarely leave much room for compromise. James found himself in the unfortunate position of having both arguments at once, often with the same adversaries.

Familiarity has dulled our sense of quite how anomalous the Church of England seemed to contemporaries in the early seventeenth century. Anglicans had broken with Rome but, uniquely among Reformed churches, had kept the old structures of church government, including an episcopacy claiming direct succession from St. Peter. The Church of England had retained, too, several of the rites and practices of Roman worship. To many English Protestants, these rites were excrescences that needed to be cleaned away. They hoped that a monarch raised in the Scottish Presbyterian tradition would, in the phrase of the time, complete the Reformation. They were to be disappointed. James, worn down by years of argument with grim Presbyterian elders, was delighted to find a united episcopacy that was both in control of its congregation and respectful of royal suzerainty. He had no more intention of humoring the Puritan radicals than he did the parliamentary radicals—often the same men.

The third dispute, over the balance of power between Crown and Parliament, was the one that historians traditionally freighted with the heaviest significance. These, after all, were the golden years of Sir Edward Coke, whose defenses of judicial and legislative independence from the interference of the executive formed the basis of Anglo-American legal teaching for the next two centuries.

King James was an enthusiast for royal absolutism. In two political treatises, *The Trew Law of Free Monarchies* (1598) and the *Basilikon*

Doron (1603), he set out the theory of the divine right of kings. No one could accuse the plain-talking Scotsman of sugarcoating his opinions: "The state of monarchy is the supremest thing upon earth, for kings are not only God's lieutenants upon earth and sit upon God's throne, but even by God himself they are called gods."

Most MPs were alarmed by the king's train of thought, and they passed frequent resolutions decrying the idea that supreme power was reposed in the Crown. We should, of course, be wary of anachronism. We know, as those MPs did not, that the quarrel over parliamentary sovereignty was to eventually lead to a civil war. There is therefore a temptation to backdate the causes of that war, to see every dispute between James and his parliamentary opponents as a step toward the confrontations of the 1640s. In fact, most MPs were more concerned with their interests as taxpayers and property owners than with their status as the people's tribunes. Nonetheless, their fiscal and religious quarrels had a way of constantly spilling over into a more general argument about the prerogatives of Parliament.

After the especially stormy session of 1614, known as the Addled Parliament, when MPs had roared unfocusedly against aspect after aspect of royal policy, James told the Spanish ambassador, "The House of Commons is a body without a head. The members give their opinion in a disorderly manner; at their meetings nothing is heard but cries, shouts and confusion. I am surprised that my ancestors should have permitted such a body to come into existence."

It is hard not to feel a smidgen of sympathy for the king. The House of Commons has always had a tendency to rowdiness, and contemporary Britons will immediately recognize the king's description from four hundred years ago. Yet James's attitude served to convince MPs that he yearned to shut down the institution that they regarded as the key defense of both their property rights and their religious freedoms. Without the House of Commons, they feared, England would become just like any Continental despotism.

Their fears were realized during the reign of James's son, Charles I (1625–49). Where James had at least had a rough bonhomie, Charles was shy and withdrawn, and never overcame his stammer. He was ill at ease with his subjects, and especially uncomfortable with their elected representatives.

The three parallel arguments that had set MPs against his father intensified under Charles. Like James, he was accused of doing too little to aid the Protestant cause abroad. The Thirty Years' War, proportionately the most lethal ever fought in Europe, had broken out in 1618. James had been in no mood to intervene, preferring to dally with Spain; and his heir had been closely associated with the pro-Spanish faction at court.

Worse, from his subjects' point of view, was Charles's marriage to a "popish queen." Two months after becoming king, Charles had been married by proxy to Henrietta Maria, daughter of Henry IV of France, before the doors of Notre Dame, rushing the procedure before Parliament could assemble, lest MPs try to forbid the banns.

Charles's marriage led to the suspicion that his children would be brought up as Catholics, and even that he had Catholic leanings himself. The first suspicion was well-founded. The second was not, but it colored the reaction to almost everything he did. Not only was his foreign policy seen as insufficiently robust; his domestic church reforms were mistrusted and resented.

Charles was closely associated with an intellectually brilliant clergyman named William Laud, who was swiftly promoted as Bishop of Bath and Wells in 1626, Bishop of London in 1628, and Archbishop of Canterbury in 1633. Laud and his party were determined to arrest the English church's slow drift toward Puritanism. They rejected the central Calvinist doctrine of predestination—the sobering idea that everyone, before birth, is already marked out for either salvation or damnation. They upheld the supremacy of bishops. Indeed, they taught that the Church of England was the *only* true catholic church, preserved by its medieval isolation from the errors and superstitions of Rome, yet also free from the heterodoxies of Geneva and Wittenberg. This was a revolutionary doctrine: like so many of the king's party, the Laudians were the true modernizers of their era.

What most upset the Puritan party were the new archbishop's attempts to impose conformity in the order of service. Laudians wanted to remove the emphasis from the sermon and restore it to the communion. It was, they believed, presumptuous for worshippers to seek to understand God. Congregations were there, rather, to adore Him, and the service should be designed to highlight the mystery of faith, with the clergyman returned to the role of mediator or intercessor. Hence the importance of beautiful surroundings, stained glass windows, candles, glorious vestments.

Some Englishmen welcomed the return of an element of mystery and ritual to their church. Others saw only an attempt to bind the country to Catholicism. Again, it is important to stress that the objections both to Laudianism and to actual Roman Catholicism were more political than theological. Charles's support for a hierarchical model of church government mirrored Laud's support for monarchical absolutism. Catholicism and autocracy were fused in the minds of contemporaries, and remained fused for nearly two centuries. Puritan congregations began to depart in droves for the "New England" across the Atlantic, seeking not total religious toleration but rather freedom from the idolatry and worldliness that they believed were corrupting Old England's church.

Even more than in James's reign, religious arguments envenomed financial disputes. The king—who, though he was devoted to his wife, was certainly no Catholic—was sensitive to parliamentary demands that he aid the Protestant cause in Europe. After all, his sister Elizabeth was married to Frederick V, the Elector Palatine, who had lost his throne after being defeated by the Catholic powers in 1620 at the Battle of the White Mountain. The trouble was that, having demanded a more belligerent foreign policy, MPs were reluctant to pay for it. They talked vaguely of piratical raids on Spanish ships that might be self-financing. They made grants to the king only on an annual basis. The war went badly, which served to inflame them the more against the king and his ministry.

Charles, meanwhile, had started to raise funds in all manner of semiconstitutional and unconstitutional ways. His lawyers revived devices that kings had used in the Middle Ages. Landowners, for example, were fined for the encroachments that their ancestors had made into royal forests since the reign of Richard I more than four centuries earlier. A forgotten edict of 1279 was found that required every individual who earned more than forty pounds a year to present himself at court and offer his services as a knight, and Charles's officials levied a fine against all those who had failed to attend his coronation in 1626. Another ancient statute, Ship Money, provided for a tax to be levied on coastal towns during war, but royal lawyers declared that there was nothing to prevent its application to inland regions.

Like other kings around Europe, Charles aimed to build up a steady revenue stream. If he were fiscally independent, as both sides well understood, he would be able to rule without Parliament.

In March 1629, he judged that his moment had come. MPs somehow got wind of a coming dissolution, and refused to heed the traditional summons to wait upon the King in the House of Lords. Two of them held the Speaker down in his chair so that the session could not be terminated, while a resolution was passed proclaiming anyone who supported the Laudian church reforms, as well as anyone who participated in collecting the King's customs duties, to be "a capital enemy to this Kingdom and Commonwealth." Indeed, those who merely paid the levies were declared "betrayers of the liberty of England, and enemies of the same." Significantly, in these short and angry resolutions, MPs branded their enemies "innovators in the Government." They were in no doubt that they were upholding the ancient constitution, while the king's party were seeking to destroy it. Later that day, Charles proved them right. Parliament was dissolved, and would not be recalled for eleven years.

The period that followed was traditionally referred to by Whigs as the Eleven Years' Tyranny, but is more primly known to historians as the Personal Rule. Charles sought to construct the kind of royal despotism that monarchs were building across the Channel and, but for the bloody-mindedness of his subjects, he might have succeeded.

The 1630s were characterized by an intensification of the religious and financial disputes. Opponents of Laudianism felt vindicated in their suspicion that church ritual and authoritarian rule went hand in hand and began to agitate for the outright abolition of bishops. Meanwhile, the king's various extralegal levies were resisted by the judiciary as well as by the now dispersed parliamentary leaders, above all John Hampden, who mounted a legal challenge against Ship Money. A few historians have argued that these years were, for most Englishmen, peaceable and stable. Taxes were low, if arbitrary, and the beautification of church buildings and religious rites accorded well with the temper of the nation.

Certainly some Englishmen took that view. They were to form the core of the king's supporters, the Cavaliers, in the coming conflict, and their ideological descendants were to become the Tory Party. Yet, as in any autocratic regime, there was also corruption, misgovernment, and cruelty. Monopolies were sold and abused. Royal favorites settled scores with old rivals. Critics of the regime were harassed and imprisoned. Englishmen began to know the anxieties, the petty humiliations, the frustrations that come with dictatorship.

Why didn't Charles succeed? What made England and Scotland defy the European trend? Partly, yet again, the natural resilience of common law, a system that had evolved to defend the liberty of the individual, and that no government could easily bend to its will. With no legislature in session, opposition to royal policy came from the Bench. Common-law judges were often remarkably fearless in striking down the king's abuses. Even when they ruled his actions to be within the law, they tended to do so contingently and with qualifications. The court that upheld the King's technical right to levy Ship Money in 1638, for example, nonetheless declared that he had no right to "impose charges upon his subjects in general, without common consent in Parliament." The presiding judge, Sir Robert Berkeley, added that "the people of the Kingdom are subjects, not slaves; freemen, not villeins to be taxed *de alto et basso.*"

As the common law persisted, so did the local substructure of legal and representative institutions. Like the Norman monarchs, the Stuarts found it much easier to make changes at court and in the church than to interfere with officeholders at shire level. Sheriffs, coroners, leetmen, justices of the peace, church wardens, and the rest carried on their work largely unmolested—as their predecessors had, in many cases, since Saxon times. These offices were often occupied by the same gentry families that filled the House of Commons, which meant that the prorogation of that assembly did not wholly shut the king's critics out of office. Remarkably, the assize courts continued to function throughout the civil war, such was the hardiness and antiquity of local institutions.

Even so, it is possible that Charles might have succeeded but for another peculiarity of Great Britain: the fact that, being an island, it needed no standing army. The King was able to rule alone only so long as he kept the country at peace. There was no way, even with all the archaic financial devices his lawyers could find, that he could afford to prosecute a war without funds from Parliament. Charles had accordingly made peace with Spain soon after the dissolution. But war was to eventually burst upon him, without warning, from the other direction.

In 1637, a market trader named Jenny Geddes set in motion the
events that would topple Charles from his throne when she hurled
a folding stool in Edinburgh's St. Giles' Cathedral. Scotland had
always had an austere approach to religion, and the king was especially keen to bring its practices into conformity with those of the
Church of England. When the dean of Edinburgh entered the cathedral one Sunday morning wearing a white surplice, there were
gasps from the congregation. As the poor man began to read from
the new royally approved prayer book, the gasps gave way to excla-
mations of "Popery!"

It was too much for Jenny Geddes. Leaping to her feet, she
screamed at the minister, "De'il gie ye colic, the wame o' ye, fause
thief! Daur ye say Mass in my lug?" ("The Devil give you colic in
your stomach, you false thief! Do you dare to say a Mass in my ear?")
Reaching behind her for the folding stool she had brought into the
service, she hurled it at the clergyman's head, whereupon some of
her fellow worshippers surged forward, yelling "A Mass! A Mass!"
and seeking to tear his surplice from his shoulders.

Rioting followed throughout the city, and soon spread across the
Lowlands. Scottish Protestants rushed to sign the Covenant, swearing to resist the popish innovations of a king who, though born in
Fife, had spent most of his life in England. The General Assembly of the Church of Scotland put itself at the head of the protests,
thereby turning them into a national rising.

King Charles responded with spectacular ineptness, raising an
army in England to reconquer his homeland. The mobilization
sparked fears on both sides of the border that, once he had a body
of troops behind him, the king might establish an altogether more
vicious regime. Anxieties were heightened by Charles's tactless inclusion of a number of Catholic officers in senior positions. Englishmen were in no mood to pay for a military adventure necessitated by
their king's obduracy. The very name of the conflict, the "Bishops'

War," tells us of its unpopularity. Plenty of English Protestants sympathized with the Scottish Presbyterians, and were horrified at the idea of paying taxes so that an unpopular king could impose religious innovations that they disliked. There was an almost universal refusal to pay the Ship Money that year.

Unsurprisingly in the circumstances, Charles lost. His soldiers were no match for the fanatical Covenanters, fighting for their faith and homeland. As the Scots advanced, the king's financial position became impossible and, with no other options available, he summoned a new parliament. The MPs who met in April 1640 had eleven years of accumulated grievances against a monarch whom many of them now regarded as a despot. They flatly refused to approve the subsidies he demanded until these grievances were remedied, and even began negotiating separately with the Scots. Horrified, Charles dissolved the so-called Short Parliament after three weeks. But his situation continued to deteriorate. The Covenanters pushed into England, occupying a number of northern cities. Eventually, the king bowed to the inevitable and, in November, summoned what was to become known as the Long Parliament, whose remnants would not be finally dissolved until 1660.

The MPs who answered that summons were acutely aware of the threat of a permanent royal dictatorship. The king tried to strike a more emollient note, promising to look into their religious and fiscal complaints, and to "reduce all matters of religion and government to what they were in the purest of Queen Elizabeth's days." But the parliamentarians no longer trusted him. They wanted revenge for the years of unconstitutional rule, and the impeachment of the royal counselors associated with it.

At the same time, they feared that Charles would, as soon as he was able, dispense with them entirely and rule by force. Some of the king's ministers were known to be seeking military assistance from Spain, while the queen was soliciting aid from her brother, Louis

XIII of France. As if these intrigues with the country's two traditional enemies were not bad enough, Charles refused to disband his
army in Ireland, and there were constant rumors that it was on the
point of embarking for London.

As Parliament met against a backdrop of demonstrations from
the London mob, a force for radicalism and Protestantism throughout the seventeenth and eighteenth centuries, the prospect of a royal
coup d'état became frighteningly real. The religious, fiscal, and constitutional arguments were pushed into the background by a more
immediate question: who was to control the army?

Any hope of compromise on that question disappeared when,
in November 1641, news reached London of a major rising by
Irish Catholics against the English- and Scottish-descended planters who had settled in the northeastern part of Ireland. As the
law then stood, only the King had the power to command troops.
But MPs were determined not to put a mass of regular soldiers at
the King's back. They declared that they would authorize no such
force unless the King placed the militia under lord lieutenants appointed by Parliament.

Their fears of a kingly putsch turned out to be well-founded.
In the New Year of 1642, Charles struck, replacing the garrison of
the Tower of London with Royalist artillerymen from the Army
of the North, announcing that Parliament would henceforward be
protected by his own troops, and ordering the lord mayor to disperse
the London mob by gunfire if necessary. On January 4, Charles rode
to Westminster at the head of three hundred soldiers and, in a mon-
strous violation of precedent and prerogative, forced his way into the
chamber and sat in the Speaker's chair. He announced that he had
come to arrest five parliamentarians who had been leading the opposition to his policies. Not seeing them present, he demanded that
the Speaker, William Lenthall, tell him where they were.

The Speaker, with great courage and dignity, replied: "May it please your Majesty, I have neither eyes to see nor tongue to speak in this place, but as the House is pleased to direct me, whose servant I am here."

The king, scanning the benches, muttered that he, too, had eyes. "Ah," he said at length. "I see the birds have flown."

The five MPs had fled to the City of London, citadel of the merchant classes who were their strongest supporters. The enormity of the king's assault on parliamentary privilege swung public opinion decisively behind them, and the House of Commons, fleeing to a temporary refuge in the livery hall of the Worshipful Company of Grocers, took a step that made armed conflict almost inevitable by assuming command of the city militia. The following week, fearing for the safety of his family, the king left London.

Events now took on a momentum of their own, as the political classes picked one side or the other. Various attempts were made to find a compromise, but trust between the two parties had long since been exhausted. Throughout May and June, a number of peers and gentry began to find their way to the king's camp at York, and there were clashes between supporters of the two sides. On August 18, Parliament declared all who supported Charles to be traitors. Four days later, in a gloriously medieval gesture, the king raised his standard at Nottingham, summoning all loyal subjects to the aid of their liege lord. The next day, it was found to have blown over in poor weather.

THE FIRST COUSINS' WAR

The conflict that followed cut indiscriminately across all the realms then inhabited by English-speakers. In Scotland, the Bishops' Wars gave way to a civil war between Covenanters and Royalists, the latter aided by an Irish army. In Ireland, there was a vicious sectarian struggle known as the Confederate War (sometimes as the Eleven Years' War), which ended with invasions from both Scotland and England and atrocities that are to this day recalled with a shudder. In England itself, there were two linked civil wars—Scots supported Parliament in the first, the King in the second—followed by an abortive attempt at a monarchical restoration. Many historians refer to these interlinked conflicts as the Wars of the Three Kingdoms (Wales was then an integral part of England), though it is perhaps more helpful to think of them as the First Anglosphere Civil War.

For there was also another place where English was spoken. Between the vastness of the Atlantic Ocean and the vastness of the interior, sparse colonies were struggling to establish themselves along the American seaboard. They rarely get a mention in histories of the civil war, largely for reasons of relative size. The population of England in the mid-seventeenth century was around five million, plus half a million in Wales; that of Ireland was perhaps two million, that of Scotland a million. The total English-speaking population of the American colonies, concentrated in New England, was no more than forty thousand, less than one-half of one percent of that of the British Isles. Yet the colonists, too, had their civil war, and ranged themselves along more or less the same lines as their compatriots across the Atlantic.

Wars are the ultimate test of nationality. Confronted by a foreign enemy, people of the same nation will quickly forget their domestic quarrels. The wars that perturbed the English-speaking world in the 1640s did not set state against state. They were not wars between Ireland and Scotland, or between England and her offshoot on the American littoral. Rather, they carved laterally across all these polities, pushing combatants into one of two camps.

In England, Scotland, America, and Ireland, there were, broadly speaking, those who fought for monarchy, aristocracy, episcopacy, hierarchy, loyalty, and land; and there were those who fought for individualism, Protestantism, representative government, and free trade. These common alignments formed the basis of the Anglosphere that was to emerge from the fighting with a shared political consciousness. They also formed the basis of the two-party system that was common to the Anglosphere from then on.

It was during the Wars of the Three Kingdoms that the words *Whig* and *Tory* were first heard, although another generation was to pass before they came into use as political labels. *Whig* is an abbreviation of *Whiggamore*, the name given to those Scottish Covenanters who opposed reaching an accommodation with King Charles. *Whiggamore* is thought to come from a Scots word meaning "drover of mares"—and thus, by implication, country bumpkin.

Tory comes from the Irish *tóraidhe*, meaning "pursued man" or, more colloquially, "outlaw." The original Tories were the defeated Irish Catholics who, dispossessed of their lands, hunted and hungry, kept up a guerrilla war from the islets of the western boglands through the 1650s. One Cromwellian officer wrote of seeking to exterminate three kinds of beast: "The first is the wolf, on whom we lay five pounds a head, and ten if it is a bitch. The second beast is a priest, on whose head we lay ten pounds. The third beast is a Tory, on whose head we lay twenty pounds."

Both terms first came into use as political insults in the 1670s. In a population where religious extremes were generally deplored, it was disparaging to liken one's opponents either to obstreperous Scottish Presbyterians or to Irish Catholic brigands. Yet, as sometimes happens, both insults were adopted with pride by their targets. They formed the basis of the two-party system throughout the Anglosphere for a century, and remained useful labels long after that. There were Whig parties in Britain and the United States until the 1860s. British and Canadian Conservatives are still sometimes called Tories.

The party labels, though, were simply the outward display of an ideological fault line that was present throughout the Anglosphere, regardless of what politicians called themselves.

Consider, to pluck an example more or less at random, George Eliot's *Middlemarch*, often said to be the greatest novel in the English language. Published in installments in 1871 and 1872, it tells the story of a provincial English city forty years earlier. It is a lengthy and spacious book, interweaving the stories of many gentry families and townspeople. It soon becomes clear to the reader that Middlemarch—which might stand for any English town of similar size—is divided into two factions, defined by social preference and churchmanship more than by political competition, though their political differences are being sharpened by the advent of the 1832 Reform Act. The words *Tory* and *Whig* appear very rarely in the novel, because the author takes it for granted that her readers will immediately recognize the schism that, in their day, too, was as much cultural as partisan. The Tory-Whig division had lasted in Britain for two centuries when Eliot was writing and, in its Conservative-Liberal form, was to last half a century more. Not until the rise of democratic socialism in the 1920s was it superseded by a different ideological disagreement. Toryism and Whiggery were so closely linked to religious denomination that children often unconsciously took in their assumptions at Sunday school. The division seemed to be, to repeat Jefferson's definition, "founded in the nature of man."

The earliest Whigs and Tories consciously identified with the Roundhead and Cavalier factions of the previous generation. (*Roundhead* was the nickname for those who had fought for Parliament against Charles I. Puritan men tended to defy the fashion for shoulder-length locks, although portraits show that most parliamentary leaders had the same luxuriant curls as their Royalist adversaries.)

That conflict had engulfed the entire English-speaking world, and the party system it gave rise to could be found across the Anglosphere.

It goes without saying that the Wars of the Three Kingdoms played out differently country by country, region by region. Everywhere the ideological struggle was modified by local rivalries. In Ireland, there was a nationalist rising by the Catholic majority against the domination of an alien ruling class. In Scotland, the war revived animosities between English-speaking Presbyterian Lowlanders and the Gaelic-speaking clans from whom they were sundered by speech, custom, and faith. Allegiances were further complicated by vendettas *within* the Highlands: some clans automatically ranged themselves against their hereditary foes. Their chiefs who, in Edinburgh, were titled in the Anglo-Norman style as earls or dukes, and talked as politicians, knew that their power rested on their ability to call a thousand claymores from the glens.

The English, Irish, Scottish, and American wars were not simply linked by the sympathies of their combatants. The pro- and anti-Stuart factions fought side by side across national boundaries. Scottish Covenanters pursued their war against the king into England, then switched sides, allying with him in return for a promise to impose Presbyterianism in England, and finally were invaded by Cromwell, who in effect annexed their country. Ireland's Catholic Confederacy sent an army to aid the Scottish Royalists, led by a Scottish clansman named Alasdair MacColla and his Irish cousin Manus O'Cahan. Having sent Royalist troops to Scotland, Ireland ended up being invaded by Puritans from both Scotland and England. Although no Irish army ever landed in England, the threat that one might land, bringing with it "popery and slavery," was arguably the single most effective piece of propaganda at the Roundheads' disposal.

In the American colonies, the factions formed along the same lines as elsewhere. Broadly speaking, Virginia was Cavalier, New England Roundhead. The main battles took place in Maryland between Catholics and Puritans. The Puritans had the better of the fighting, though Lord Baltimore and his faction were restored to power when Charles II regained his father's throne in 1660.

What were the dividing lines? One aspect was social. Cavaliers could draw on the support of most of the aristocracy and their loyal tenantry. Roundheads were more popular among the merchant classes, and the City of London proved to be their most valuable resource.

A second aspect was geographical. In England, Parliament drew its support from the southeast and, above all, from was what became known as the Eastern Association: the counties of Norfolk, Suffolk, Essex, Cambridgeshire, Huntingdonshire, Hertfordshire, and Lincolnshire. It was from these flat and fertile lands that Oliver Cromwell drew his psalm-singing cavalrymen. The Cavaliers won their chief support from the hillier and wetter lands of northern England, the West Country, and Wales.

In Scotland, the Royalist heartland was in the largely Catholic Highlands (though with the important exception of Clan Campbell lands) and the largely Episcopalian northeast. In Ireland, the king's supporters dominated most of the island except for the Protestant settlements of the northeast and some of the eastern coastal towns.

Americans followed the pattern of their ancestral regional loyalties. New England had largely been populated by settlers from the counties of the Eastern Association, who had named their new towns for their old: Boston, Billerica, Cambridge, Dedham, Hatfield, Harwich, Ipswich. They felt close to their Puritan kin in England—kin from whom they were separated by no more than one or two generations. When the fighting started, Massachusetts Puritans began streaming back across the Atlantic to fight alongside their cousins.

More than half of all Harvard graduates in the 1640s saw action on the Roundhead side in the English Civil War. By contrast, most Virginians, being Anglicans and admirers of England's great rural estates, which they were already starting to replicate, remained loyal to the king.

Underpinning the social and regional differences was the confessional schism. Because, in our own age, we have largely overcome sectarianism, we can be tempted to downplay this aspect of the war, instead focusing on its class and geographical factors. But the truth is that all the other factors were tinged with religion.

Puritanism was stronger in the towns than in the countryside; stronger among artisans and merchants than among landowners; stronger in southern and eastern England than in northern or western; stronger in the Scottish Lowlands than in the Highlands; stronger in New England than in Virginia; strong in Ulster but unknown in the rest of Ireland. It seems also to have been stronger among the older generation than the younger. Again, we need to remind ourselves that the divine right of kings was a fashionable new idea.

The story of the war itself can be briefly told, for this is not a military history. In retrospect, we can see that King Charles's only chance of victory was to capture London at the outset. His failure left the Roundheads with overwhelming advantages of finance and population, which told over time.

The war was, at first, fought with remarkable moderation. There were few abuses, and almost no atrocities. When one side or another seized territory, its prominent opponents usually had nothing worse to fear than house arrest. Sometimes they were simply asked to give their parole, promising not to bear arms again, and allowed to go free.

We need only look at what was happening in contemporary Germany to see how extraordinary this was. The Thirty Years' War, which was then entering its bloody final spasm, was punctuated by abominations: the casual slaughter of civilians, the murder of

prisoners, the burning of homes. Fifteen hundred towns and eighteen thousand villages were completely razed, and those that survived could take decades to recover. Germany and the Czech lands
lost perhaps a third of all their inhabitants, and the depopulation
was even more severe in the intense war zones. Half of all Branden-
burgers perished, three-quarters of all Württembergers.

One reason for the relative temperance of the English Civil War
is that the overwhelming majority of Englishmen were Anglicans.
Their religious differences, in other words, were contained within a
single church. Four in five Englishmen felt comfortable within the
Church of England. A tiny number of Catholics—sixty thousand in
1640—practiced their faith openly, paying the fines for recusancy.
A few more, the so-called Church Papists, did so privately, while attending Anglican services so as to avoid sanction. The total number
is hard to estimate, but was probably less than 5 percent of the population, although disproportionately weighted toward the upper
classes. The number of Puritans was higher: perhaps 15 percent.

During the civil war, some Catholics kept their heads down,
while others fought for the king, who received generous financial
donations from Catholic aristocrats. Almost all Puritans backed
Parliament. But the two religious extremes very rarely came into
direct conflict. Both armies were led and largely officered by Church
of England men.

In England, Puritan and Catholic soldiers tended to come up
against Anglicans rather than against each other (Ireland was a different story, as we shall see). But, on the exceptional occasions when
they clashed directly, they displayed all the ferocity shown by their
coreligionists on the grim battlefields of Germany.

Bolton was a Puritan cloth town surrounded by largely Catholic
Lancashire countryside. When a Royalist force, led by Prince Rupert
and containing many Catholics, stormed the town on May 28, 1644,
at night and in heavy rain, several hundred defenders and civilians

were massacred. The following year, Basing House, whose owner, the Marquess of Winchester, had turned it into a refuge for Catholics, including several priests and Jesuits, was taken by Cromwell, who had ignored the defenders' request for a parley. Again civilians were slaughtered alongside soldiers. (One of those taken alive, naked and wrapped in a blanket, was the architect Inigo Jones, whose gorgeous Italianate works, including the Banqueting House on Whitehall, had served to convince Puritans that the king was aesthetically, if not technically, a Roman Catholic.)

These clashes stood out precisely because they were so shocking. Although the shires most affected by the war suffered from requisitioning and plunder, there was nothing like the savagery that, in other countries, was considered normal in war. Some Roundheads began to suspect that their leaders, the Earls of Essex and Manchester, were pulling their punches. Again and again, parliamentary forces seemed reluctant to prosecute the war with the requisite determination. The aristocratic generals hoped for a negotiated peace, which would see the king return as a constitutional monarch. They did not want to behave in a way that might forestall such a settlement. As Manchester dolefully put it, "If we beat the King ninety and nine times yet he is King still, and so will his posterity be after him; but if the King beat us once, we shall be all hanged, and our posterity be made slaves."

The mood of his troops, however, was far less compromising. The Puritans, fired with religious zeal, wanted total victory and a new Jerusalem. Eventually, they pushed through the Self-Denying Ordinance, obliging all parliamentarians—including Essex and Manchester—to resign their commissions. In 1645, leadership passed to Sir Thomas Fairfax and to a rising cavalry commander from the fens called Oliver Cromwell, whom the Puritan troopers recognized as one of their own: a man who could sing and psalm with them. The old county militias were reorganized into a tough, professional force, the New Model Army. From then on, it was only a matter of time. In June 1645, two Royalist armies were crushed at Naseby in Northamptonshire, and the king's cause was lost.

The Parliament men sought to negotiate new terms with their defeated monarch, but it soon became clear that, even while he was haggling with his captors, Charles was covertly intriguing to recover his ascendancy. Eventually, he contracted a secret bargain with the Scottish Covenanters to impose Presbyterianism in England in exchange for their assistance. There were Royalist risings around England, and a Scottish army crossed the border in their support. Cromwell crushed his combined enemies at the Battle of Preston in 1648, and the second English Civil War came to an end.

Whereas the first war had ended in a spirit of magnanimity and reconciliation, this time there was a desire for vengeance. The king had confirmed his detractors' worst fears, proving treacherous, duplicitous, and power-hungry. The logic of Manchester's complaint was inexorable. As long as Charles breathed, there could be no peace, no permanent settlement, no return to a balanced constitution. Army leaders and MPs were driven to a conclusion that, eight years earlier, hardly a man in England would have ventured: the king must die.

On January 30, 1649, having requested a second shirt so that his subjects should not take his trembling for fear, the king was decapitated outside Inigo Jones's Banqueting House. As the contemporary poet Andrew Marvell put it, "He nothing common did or mean upon that memorable scene." Charles's dignity in death led to a cult of martyrdom that was, eleven years later, to ease his son's restoration.

The wars that had awakened cross-border affinities among the English-speaking peoples now ended with the effective collapsing of those borders. For the first time, the entire Anglosphere—that is, the British Isles plus the American colonies—came under a single regime, as Oliver Cromwell imposed his iron rule. Between 1653

and 1659, a unitary government ruled a state known as the Commonwealth of England, Scotland, and Ireland. The chief enemy of that government was not secessionism in any of the constituent realms, but a Royalist resentment that smoldered in parts of all of them—and which even flickered feebly across the Atlantic.

In 1652, a fleet was sent to bring Virginia, whose governor had immediately recognized Charles II as king following his father's death, back into the fold. Heavily outnumbered on their own continent, the Virginian Royalists complied without a shot being fired in anger. Ireland, where the vast majority of the population had been at least nominally allied to the king, was an altogether bloodier affair, and the name of one especially brutal engagement echoes to our own day.

In September 1653, Cromwell besieged the Irish town of Drogheda, which was held by a largely Catholic garrison. He invited the governor, an English Royalist, to surrender. The governor, knowing that there were four thousand Royalist troops nearby under the Duke of Ormonde, refused. Cromwell proceeded to take the town by force, with heavy losses. Once inside, the parliamentary forces slaughtered every soldier they could find. Several civilians died, churches were sacked, priests clubbed to death. At least two Royalist officers who had been taken prisoner were later shot in cold blood.

Cromwell's behavior shocked many in England, and is even now remembered and resented in Ireland. By the standards of its time, it was not unusual. Contemporaries well understood that, when a town was besieged, the garrison inside held the advantage. The investing soldiers were prey to hunger, disease, and attack from two directions. The convention was that, if the defenders refused to surrender, they might, under the laws of war, be put to death. In the wars then under way in Europe, such massacres were routine.

In the Anglosphere, however, with the partial exceptions of the sieges of Bolton and Basing House already discussed, wholesale killings were at that time unknown. The sectarian explanation is the obvious one. Without the moderating influence of middle-way Anglicanism, English-speakers were as wont to slaughter one another as any European religious fanatics. Justifying his behavior at Drogheda in a letter to Speaker Lenthall, Cromwell claimed divine sanction. "I am persuaded that this is a righteous judgment of God on these barbarous wretches, who have imbrued their hands with so much innocent blood; and that it will tend to prevent the effusion of blood for the future, which are satisfactory grounds for such actions which cannot otherwise but work remorse and regret."

Again and again, we are struck by the aggressive nationalism of the Roundheads—a nationalism that transcended England, and extended to the entire English-speaking population. Supporters of Parliament, foreshadowing the complaint in the U.S. Declaration of Independence about "foreign mercenaries," constantly accused the Royalists of seeking foreign aid. They had a point. The king solicited help from Holland, Denmark, Spain, France, and the Vatican. His pleas were unsuccessful: no Continental prince would help a man known to be untrustworthy and known, too, to be petitioning their rivals. The king's foreign policy was a disaster, for, as the majestic Whig historian G. M. Trevelyan put it in 1924, Charles was asking for "military aid which his detractors could always announce to be coming, but which never came. The King was defeated by these phantom armies which he himself invoked."

Yet, though they raged convincingly against these overseas phantoms, English Roundheads had no qualms about inviting Scots to fight alongside them, and treated supporters of the Irish Catholic Confederacy not as foreign foes, but as rebels. This is not to say that Anglosphere identity had supplanted English, Scottish, or Irish loyalties. But there was now an English-speaking matrix, whose peoples were not wholly foreign to one another in the way that a Hungarian or a Swede might be.

This was a key moment in the development of an Anglosphere identity based on political principles. The English-speaking peoples began to believe that the things that set them apart from Europe—personal freedom, the rule of law, representative government, and the rest—were common properties of the Anglophone world. Presbyterian ministers preached sermons likening England and Scotland to Israel and Judah, kindred nations under God's special care. Ulster Protestants and New Englanders had an even stronger sense of being part of a providential union of chosen peoples.

While this sense contained elements of chauvinism, there were also respectable grounds for believing that the outcome of the civil war had secured the Anglosphere for the cause of liberty—liberty being understood to be inseparable from Protestantism. Trevelyan invited his readers to suppose that the war had ended differently: "The current of European thought and practice, running hard towards despotism, would have caught England into the stream. England would have become a mere portion of the state system of Europe, had she not, by the campaign of Naseby, acquired her independent position between the old world and the new."

The republican interlude saw an explosion of radical ideas, some eccentric, some sublime. Pamphleteers and preachers demanded equality for women, full democracy, the abolition of censorship. Greatest among them—indeed, arguably the greatest English-language writer after Shakespeare—was John Milton, a Cromwellian civil servant who advocated, among other things, divorce and free speech.

Milton is a demigod in the pantheon of Anglosphere libertarianism. In the words of the Victorian author and Anglican clergyman Mark Pattison:

> He defended religious liberty against the Prelates, civil liberty against the Crown, the liberty of the Press against the executive, liberty of conscience against the Presbyterians, and domestic liberty

against the tyranny of canon law. Milton's pamphlets might have
been stamped with the motto which Selden inscribed in Greek in
all his books, "Liberty before everything."

Liberty, in Milton's mind, did not mean an absence of rules: this
he called "license," and heartily disliked. Liberty, rather, meant the
freedom that comes from the virtuous and informed exercise of independent judgment. Such virtue might exist only if there were alternative points of view. From the clash of competing ideas, Milton
believed, truth would emerge: "Where there is much desire to learn,
here of necessity will be much arguing, much writing, many opinions; for opinion in good men is but knowledge in the making."

Milton's political ideas colored his greatest work, *Paradise Lost*,
a retelling of Adam's Fall intended "to justify the ways of God to
man." It is impossible to read his exquisite verses without being
struck by how sympathetically Milton treats Adam and even Luci-
fer. God, by contrast, comes across as arid, lordly, and cruel. Milton
heartily disliked authority and, just as he railed against the claims of
popes and princes, so he allowed his libertarian radicalism to tinge
even his view of the Almighty.

One happy consequence of the interregnum was a religious pluralism then unknown in Europe. Plenty of countries practiced religious *toleration*, in the sense of allowing their minorities to worship
as they pleased. What was unheard-of was to allow different faiths
to proselytize freely.

A combination of religious pluralism and a Puritan emphasis on
the Old Testament prompted a further happy consequence. In 1655,
a Dutch Jew named Menasseh Ben Israel approached Cromwell
with a proposal to readmit his coreligionists, who had been expelled
from England in 1290. The old cavalry commander, Philo-Semitic
by temperament and keen on promoting trade with the Netherlands,

readily agreed. In consequence, Jews were able to settle in England with few of the legal disabilities that they suffered on the Continent, and their status was little altered after the Restoration. As Paul Johnson showed in his monumental *History of the Jews*, England was, before the birth of the United States, the best place to live and practice as a Jew. The reason was that, elsewhere, Jews had been placed in a separate legal category in the days when ecclesiastical courts claimed no jurisdiction over non-Christians. This separate status made European Jews vulnerable down the centuries to all manner of discrimination and persecution. In England, by contrast, Jews were subject only to the relatively mild restrictions placed on all non-Anglicans. A Jewish Briton in the eighteenth century was in the same legal category as a Methodist or Catholic Briton (all restrictions were lifted in the nineteenth century).

It may seem fanciful to claim Philo-Semitism as a Whig virtue, but the association is of ancient standing, and occasionally expresses itself in the most unexpected ways. As late as the 1990s, during the final phase of the violence in Northern Ireland, you could tell the republican areas from the loyalist by the flags. Alongside Irish tricolors, the republicans would fly PLO banners. The loyalists, heirs to the ancient Whig and Williamite cause, responded by hoisting Israeli Stars of David next to their Union flags.

While the Commonwealth saw an expansion of liberty, it soon became clear that there would be no commensurate expansion of democracy. The victory of the parliamentary armies did not mean a victory for parliamentary government. On the contrary, Cromwell lapsed into military dictatorship, dismissing what was left of the House of Commons and ruling the Three Kingdoms through a coalition of preachers and pikemen. England, Scotland, and Ireland were placed under military government—as America would doubtless have been, too, had there been the slightest need. In the event, though, the colonists were among the strongest supporters of the Commonwealth and sheltered many of its veterans after the Restoration.

The traditional governing classes opted out of politics and withdrew to their rural estates. Cromwell made various attempts to entice them back, even summoning a successor to the House of Lords called, with stunning banality, the Other House. But the old peerage politely declined to have anything to do with the man now styled Lord Protector.

While he lived, Cromwell held the country together by the force of his personality. When he died, as in so many military dictatorships, his son was put up as a figurehead, while the regime's senior officials and generals jockeyed for position. The truth was that, as one historian put it, "Oliver ruled England from his urn." It quickly became clear that no successor could command the necessary loyalty. The only regime that would be seen as legitimate was a restoration of "the ancient constitution"—that is, the two old Houses of Parliament and the monarchy. It was just a matter of time before one or another general swam with the current.

In the event, the general was George Monck, the military governor of Scotland. As his army marched on London, it became clear that no one had the will to oppose him. The residue of the old 1640 House of Commons was summoned and promptly voted to dissolve itself and call fresh elections. A new Parliament met, this time comprising both chambers. One of its first acts was to receive a conciliatory letter from the late king's exiled son, promising a general pardon, payment of army arrears, the confirmation of property rights, and, with some qualifications, religious pluralism. The prince, very properly and modestly, made all these pledges subject to parliamentary approval. The new Parliament, less Puritan and less radical than its predecessor—partly because its members were from a younger generation—voted to efface all the legislation of the Interregnum and bring back the monarchy. Charles II was recalled from France and reached London on May 29, 1660—his thirtieth birthday.

FROM RESTORATION TO REVOLUTION

The Royalists, now with a solid parliamentary majority behind them, had a handful of scores to settle. The surviving regicides—that is, the men directly connected with Charles I's trial and execution— were tried and, in most cases, put to death. Those already deceased had their corpses dug up and mutilated. Dispossessed Cavalier families got their lands back.

The Puritanism (both upper- and lower-case) of the Cromwellian regime was swept away: theaters reopened, and taverns and brothels enjoyed a boom, fashionable gentlemen taking their cue from their priapic king.

The Church of England was finally given a uniformity of structure and service, prompting many Puritans to leave it. The authority of bishops was confirmed, many Laudian innovations were formalized—Laud, a mild and devout man in his personal life, had been executed by order of the parliamentary authorities in an act of vindictiveness that reflected badly on its authors—and the official theology drifted away from Calvinism and toward an emphasis on free will (not that theology has ever mattered much to Anglicans, whose church spans a wide range of beliefs).

The new ecclesiastical settlement was, however, important in a different way, less noticed by historians: it wrenched control of the Church of England away from the Crown and placed it firmly within the jurisdiction of Parliament—a reform that had been resisted by every previous monarch.

Once again, it is as important to notice what contemporaries took for granted as what they felt the need to state. Historians understandably emphasize the monarchist and High Church sympathies of Charles's "Cavalier Parliament," evident from various bills and resolutions. Yet several things went without saying. There was no longer any question that the House of Commons had a monopoly of the right to raise revenue through taxation. It was unthinkable that any monarch would again prejudice the right of free speech in Parliament, let alone storm the chamber as Charles I had done. (To this day, when a court official known as Black Rod comes to summon MPs to hear the monarch's formal opening of Parliament, the door to the chamber is ritually slammed in his face.) The Triennial Act now guaranteed regular elections—though Charles II found various technical ways to get around its provisions.

The Restoration of 1660 had been a restoration, first, of the legitimate Parliament, and only second of the monarchy. The trouble was that the new king didn't see it that way. He had experienced an altogether grander and more splendid form of monarchy during his time as a refugee in France. Like his grandfather and father, he wanted more money than Parliament was ready to grant. Unlike them, he had no way of raising revenue extralegally.

His solution was to turn to his cousin, Louis XIV. It is hard to think of a less popular course of action. France was England's traditional foe. Charles's own years of exile there, his French mother, and his swarthy looks combined to make him subject to the anti-French prejudices in England, which, as we saw in chapter 3, dated right back to the Norman Conquest. France was at that time the supreme Catholic power in the world, seen as a threat to the European balance by every other nation, but especially by Dutch Protestants, who feared for their independence. And, of course, Louis himself was the supreme exponent and exemplar of monarchical absolutism. Even keen Royalists in England and Scotland flinched a little from the dazzle of Le Roi Soleil.

Charles, though, was greedy for cash, and had no more love of haggling with Parliament than had any of his dynasty. He was quite happy to buy a quiet life with Louis's gold. In 1662, he sold Dunkirk, captured by Cromwellian troops four years earlier, for £320,000—horrifying most of his countrymen who, quite apart from considerations of wounded patriotism, feared that the city would become a privateering base against English shipping, as indeed it did.

The sharpness of the public reaction drove Charles to undertake subsequent negotiations clandestinely. In 1670, he negotiated the Secret Treaty of Dover with Louis. In exchange for a subsidy of two million crowns, Charles agreed to abandon the Triple Alliance (with Sweden and the United Netherlands) and join Louis in an attack on the Dutch. He also promised the French monarch to make a public profession of Catholicism as soon as an opportune moment presented itself—a promise he eventually kept on his deathbed.

It is difficult to think of a more unpatriotic action by any British leader. Here was Charles selling himself to his country's chief enemy, betraying an ally, seeking to evade Parliament, and doing the whole thing furtively.

The confidential clauses of the treaty did not come to light for another century, but they would not especially have surprised Charles's subjects. His Francophilia was no secret, his impatience with Parliament visible to all, and his admiration for Continental absolutism undisguised. Even his Catholic sympathies were widely suspected.

Indeed, the curious thing is that his MPs were so much more restrained than their fathers had been during his father's reign. Opposition to Charles II was moderate, constitutional, and parliamentary. Only during his final years was there any direct resistance, and it came from a tiny handful of republican hotheads. What had happened to the Good Old Cause, which, during the Commonwealth, had carried every enemy before it?

Part of the answer is that English Puritanism had become fee-bler, its life-force drained by waves of emigration to America. But the larger explanation is that no one much cared for the heir apparent.

"Don't worry, Jimmy," Charles is supposed to have replied when his brother chided him over his lack of protection, "they'll not kill me to make you king." The story may be apocryphal, but the senti-ment it expressed was real. Charles had at least seventeen children, from eight mistresses (he acknowledged twelve of the offspring); but every one of them was illegitimate. His wife, the unhappy Portu-guese princess Catherine of Braganza, credited with popularizing tea-drinking in England, had three miscarriages and bore him no heir. The throne was therefore due to pass on Charles's death to his younger brother, James, Duke of York.

In 1673, in a mood of almost hysterical paranoia about Catholic conspiracies, Parliament passed the Test Act, requiring all office-holders to swear an oath denying aspects of Catholic doctrine. James resigned as high admiral rather than comply, thereby advertising to all the faith he had practiced privately for some years.

The rest of Charles's reign was dominated by attempts to change the succession rules so that someone other than James should be next in line. It was now that the Whig and Tory factions came into being. Whigs wanted to exclude James II, fearing that a Catholic monarch would impose a French- or Spanish-style despotism upon them. Tories, while generally no lovers of Catholicism, believed in hereditary right, and balked at the idea that an assembly of mortals could tamper with succession rules they saw as divinely sanctioned. Although the struggle was, on one level, sectarian, it turned on a deeper question: was sovereignty vested in the King or in Parliament?

It is worth observing that political parties—that is, parties with a set of beliefs that went beyond being in favor of or against a partic-ular regime—were to remain almost unknown in most of Europe for another century and more. Even today, the solidity of the party system in the Anglosphere is unusual. In many countries, including several in Europe, it is a rare party that outlives its founder. Yet, in 1670s London, Tories and Whigs were behaving much as political parties have ever since, differentiating themselves by small signs, wearing different-colored ribbons, plotting in smoke-filled rooms: Whigs in coffeehouses, Tories in alehouses.

The Tories had the best of it, in the sense that the various attempts to get an Exclusion Act through Parliament failed. But matters deteriorated almost from the moment James II acceded to the throne in 1685. There was an established convention that a new monarch should summon a fresh Parliament. But James hesitated for reasons that still angered Lord Macaulay, greatest of all the Whig historians, more than a century and a half later:

> The moment was, indeed most auspicious for a general election. Never since the accession of the House of Stuart had the constituent bodies been so favorably disposed towards the Court. But the new sovereign's mind was haunted by an apprehension not to be mentioned, even at this distance of time, without shame and indignation. He was afraid that by summoning his Parliament he might incur the displeasure of the King of France.

The resentment that English-speakers felt toward the French despot was heartily reciprocated. Louis didn't want a free parliament across the Channel giving his own subjects ideas. And in James, he had an even surer ally—or, as Macaulay put it, a "vassal and hireling"—than in Charles.

Not that Louis was at all helpful to his cousin. In 1685, he revoked the Edict of Nantes, under which a measure of toleration had been granted to French Protestants. The long-term result of this decision was to tip the global balance of power permanently from France to the Anglosphere, as hundreds of thousands of the most enterprising and mercantile French citizens fled to Great Britain, North America, and South Africa—among them Paul Revere's ancestors. The short-term impact was to revive all the fears in Great Britain, never far beneath the surface, of persecution at the hands of a Catholic monarchy. With a modicum of sensitivity, James might have soothed these fears. Instead he did everything he could to inflame them

Generations of subsequent historians have sought, almost on principle, to overturn Macaulay's damning verdict on James II, but none has really succeeded.

James II was perhaps the only Stuart without any redeeming virtue. James I had been coarse, vulgar, and mercurial, but also clever and affable. Charles I was treacherous to his friends and vindictive to his enemies, yet was at least physically brave. Charles II was idle, sybaritic. and arrogant, but could, when necessary, be charming and shrewd. In James II, it is hard to find any quality at all. He was dull-witted, sly, bigoted, self-pitying, humorless, obstinate, and cowardly. There might be said to be some virtue in the constancy with which he clung to his religion, but his ham-fisted attempts to impose it on others were disastrous for his Catholic subjects.

The Duke of Lauderdale, a senior minister under Charles II, delivered a perceptive summing-up of the future king in 1679: "The prince has all the weakness of his father without his strength. He loves, as he saith, to be served in his own way, and he is as very a papist as the pope himself, which will be his ruin."

So it proved, and more quickly than anyone had guessed. Few of James's subjects wanted to risk another civil war. The rule of the major generals had discredited the antimonarchist cause, and the official doctrine of the Church of England was obedience to the sovereign. When the House of Commons was eventually allowed to meet, it voted the new king a generous subsidy and seemed ready to repeal various anti-Catholic restrictions: few MPs questioned the king's right to ease the lot of his coreligionists.

But James lacked any notion of either tact or tactics. He immediately began to promote Catholics in the administration, in Oxford and Cambridge colleges, and—most alarmingly for his subjects—in the army. He demanded that Parliament move faster on scrapping the Test Act, yet wanted to tighten the restrictions on Nonconformist Protestants. He pocketed every concession charmlessly, while all the time demanding more.

As hostility mounted, James began to purge every independent institution in his kingdom: the City of London, the lord lieutenancies, the universities, the municipal corporations, the magistracy. The king's opponents understood that he was embarked on an unconstitutional power grab that could not be resisted by constitutional means alone. They clustered around his son-in-law, William of Orange, champion of the Protestant cause in Europe. An exiled Whig opposition in the Dutch Republic began to explore ways to replace the king.

Eventually, even James realized that he had run out of allies and, in a late attempt to broaden his appeal, he turned 180 degrees and offered to lift all the penal laws against Protestant Dissenters, too. It was too late. No one trusted him anymore, neither the Anglicans, nor the Dissenters, nor indeed, with the exception of a handful of Jesuits and young fanatics, the Catholics, who could see perfectly well where his policy was leading them.

The king ordered that his edict promising religious equality be read out in every church. A number of Church of England clergymen refused, believing that the measure threatened the disestablishment of their church, and that the king's true intention was to restore Catholicism at musket point. Opposition was led by the Archbishop of Canterbury and six of his bishops. For the king to have fallen out with these mild men, who believed in their bones that their duty was passive obedience to the supreme governor of their church, is extraordinary testimony to James's character flaws.

The Seven Bishops, as they became known, petitioned the king in the gentlest of language, begging to be excused from having to read out his text. The king's almost Caligulan response was to have them charged with seditious libel and committed to the Tower of London. Never before or since have Anglican bishops enjoyed such universal acclaim in England. Vast crowds cheered the elderly prelates on their way to the Tower of London, whose guards knelt for a blessing as they passed. When their case came before the Court of the King's Bench in June 1688, they were all acquitted: yet another tribute to the common-law system that was freedom's surest ally during the Stuart years.

While the bishops were awaiting trial, something happened that made the king's deposition certain: on June 10, 1688, his wife gave birth to a baby boy. Until that moment, the heir presumptive had been Princess Mary, daughter of his first marriage, who had been raised a Protestant. It was widely assumed that she would not rule alone, but would hand the reins of government to her husband, William of Orange, the quasi-hereditary stadtholder of the Dutch Republic.

James's second wife was Mary Beatrice of Modena, a zealous Catholic whose family had traditionally been French clients. Mary Beatrice had been delivered of three stillborn children and was thought unlikely to conceive again. There were rumors, both about her infertility and about the king having become sterile following a dose of venereal disease in the 1660s. (James was almost as libidinous as his brother Charles and, in later life, saw the loss of his throne as divine punishment for his adulteries.)

When the prince, also called James, was born, much of the country was incredulous. It seemed altogether too convenient that, just when the Jesuits wanted a male Catholic heir, one should be magicked up. It was alleged that the queen's pregnancy had been a hoax, and that a stray brat had been smuggled into her chamber in a warming pan (a long-handled metal pot containing hot embers, used to warm beds).

The story was preposterous. The birth of a royal heir was a public affair in those days, and queens gave birth in the presence not only of their ladies-in-waiting but also of several senior male court officials. Yet the warming-pan theory was widely believed, then and later. Even William of Orange seems seriously to have entertained it, and the two stated justifications for his invasion a few months later were to summon a free Parliament and to hold a public inquiry into the circumstances of the infant's birth.

Not that he needed much justification. The birth of the little boy—James III to his supporters, the Old Pretender to his opponents—had changed the situation utterly. As long as the Protestant succession looked secure, most Tories and some Whigs were prepared to put up with James's oppressions and cruelties. The king was in his fifties, and even he understood that his elder daughter and her husband would eventually reign as Protestants. But the prospect of an enduring Catholic dynasty altered everything. Now James had every reason to pursue what was almost universally believed to be his design for the forcible reconversion of his realms and the transformation of his kingly office into a French-style imperium.

Whigs and most Tories banded together to seek regime change. A number of aristocrats wrote formally to William offering him the throne. Risings were planned around the country, and William readied an invasion fleet so large its spreading sails spanned the breadth of the English Channel, so that one wing saluted Dover while the other saluted Calais. William landed in Devon on that most auspicious of Protestant dates: November 5.

With the north and the Midlands in revolt, James had few options. Staying in London while his enemies encircled him would give the impression that he had already conceded. On the other hand, advancing west to meet the Dutch would risk a revolt in the capital, whose inhabitants were always ready to riot in support of Parliament and Protestantism. James split the difference, advancing halfway, hemorrhaging support with every step, and encamping at Salisbury.

There was a minor clash in Somerset—some fifteen men died in total—from which James's troops came off the worse. As his soldiers began to desert in droves to William, the king got a nosebleed, which he interpreted as a sign of divine disapprobation. He seems to have decided at that moment to abscond to France, although there was a second skirmish at Reading before he fled. Incompetent to the last, he managed to get himself captured by some Kentish fishermen and returned briefly to London, before being allowed by William to slip through the Dutch lines and bolt again, this time for good.

James's cowardice—or superstition, to put the kindest possible gloss on it—convinced his subjects that the change of regime had been God's handiwork. To overthrow a monarch without war— almost without bloodshed—was seen as miraculous. Certainly the settlement that followed constituted a small secular miracle. It ensured that the Anglosphere developed along capitalist-democratic rather than centralist-dirigiste lines. It remains the closest thing the United Kingdom has to a written constitution. In the United States, its principles directly and immediately inspired the Constitution. We call it the Glorious Revolution for good reasons.

A CROWNED REPUBLIC

There was nothing new about toppling a king. It had happened many times before, by insurrection, by conquest, by palace coup, by assassination. Being a monarch was one of the most dangerous occupations of its era. Shakespeare has the deposed Richard II complain lyrically about a king's lot:

For God's sake, let us sit upon the ground
And tell sad stories of the death of kings;
How some have been deposed; some slain in war,
Some haunted by the ghosts they have deposed;
Some poison'd by their wives: some sleeping kill'd;
All murder'd: for within the hollow crown
That rounds the mortal temples of a king
Keeps Death his court.

What happened during the Glorious Revolution was different. This was not a backstairs murder, like that which had removed Richard II; nor yet a kangaroo court, like that which had condemned Charles I. James II was deposed by the solemn decision of a full and legitimate Parliament.

Ever since the Glorious Revolution, it has been tacitly understood that the British people might hire and fire their kings just as they might hire and fire their MPs. This astonishing fact has been smothered under layers of pomp and tradition. Many of these coverings are still in place: the coaches and scepters, the military and church ceremonies, privy councillorships and knighthoods, the language used on formal occasions by "Her Majesty's Government." A casual observer might assume that the sixteen Anglosphere and Commonwealth states that recognize Elizabeth II as head of state are less than fully democratic. Yet no one seriously disputes the rights of their legislatures to choose any head of state they please. When, in 2012, they agreed jointly to alter the succession law to remove the bias against daughters, there was no demurral.

The Glorious Revolution was the moment when the Anglosphere took off, developing into a small-government, individualist, mercantile state-system. It was now that the English-speaking peoples began to look outward, building great navies rather than land armies, replacing the old guilds and monopolies with modern trading houses, creating global markets. Nor can these developments, by any stretch of the imagination, be seen as part of a general European trend. At the end of the seventeenth century, most European states were moving toward what historians now call "enlightened despotism"—and most of the rest of the world was stuck with despotism, pure and simple.

It is again worth stressing that, in its contemporary usage, the word *revolution* meant a full turn of the wheel, a restoration of that which had been placed on its head. The "revolutionaries" themselves were clear about what they were doing, and why. During his flight, James had tossed the Great Seal, without which no Parliament might legally be summoned, into the Thames. So a representative group of MPs and peers formally asked William of Orange to summon a convention "for the preservation of our religion, rights, laws, liberty and property," and "the establishment of these things upon such sure and legal foundations that they may not be in danger of being again subverted."

The fact that James had run away made their task much easier. While most Whigs would have been happy to exclude him in any event, many Tories, especially in the House of Lords, were attached to the principle of legitimate succession—the principle, after all, to which they owed their own estates and titles. In the event, they were able to declare that James had "abdicated the government, and that the throne is thereby become vacant."

Believing, or affecting to believe, the warming-pan farrago, they offered the throne to James's eldest daughter, Mary, who made clear that she would accept it only if it were shared with her husband. The accession of the eldest recognized child might superficially look like a nod toward the principle of legitimate succession, but it was no such thing. Parliament laid down that the Crown would afterward pass to Mary's younger sister, Anne, rather than to any family of William's. MPs barred Catholics and their spouses from the throne.

Subsequent legislation decreed that the House of Commons should continue in session for six months after a monarch's death, rather than automatically dissolving and waiting for the new king to call fresh elections—the end of any remaining pretense that Parliament depended on the monarch rather than the other way around.

The contractual nature of monarchy had, as discussed in chapter 2, long been implicit in England's Coronation Oath. Now, it was made explicit. Whereas James and his predecessors had promised to "confirm to the People of England the Laws and Customs to them granted, agreeable to the Prerogative of the Kings thereof, and the ancient Customs of the Realm," William and Mary and their successors swore "to govern the People of this Kingdom, and the Dominions thereunto belonging, according to the Statutes in Parliament agreed on and the Laws and Customs of the same."

The Anglosphere was doubly lucky in its new king. For one thing, William's main interest was in bringing England into his war against Louis XIV, not in its domestic politics. He was happy to let Parliament take the lead at home provided he got what he wanted abroad.

From William's point of view, that trade-off was a roaring success: England (later Great Britain, later the United Kingdom) embarked on a series of wars against France that were to last, on and off, until 1815, and the interludes between were characterized by espionage and intense commercial rivalry. The empowerment of Parliament resulted in an intensification of the anti-French foreign policy that William had wanted, and ensured that it lasted well beyond his lifetime. Some historians refer to the period as the Second Hundred Years' War. Britain fought France between 1689 and 1697, 1702 and 1713, 1743 and 1748, 1756 and 1763, 1778 and 1783, 1793 and 1802, and 1803 and 1815. Those wars, and the casting of France as a semi-permanent enemy, encouraged people from every part of the British Isles to stress the things that made them different from the French:

their parliamentary system, their common law, their Protestantism, their attachment to personal freedom.

English-speakers were fortunate, too, in their new king's background. The Dutch Republic, like the Anglosphere, valued property rights, free trade, and limited government—though it, too, was at this time oligarchic rather than democratic in its government. In one sense, the Dutch were ahead of the English-speakers: they had already developed a recognizably capitalist economy, based on joint stock ventures and limited liability. The oceangoing Hollanders, indeed, were the only other people on earth who were evolving a libertarian-democratic model of government. Not that this evolution was coincidental. During the seventeenth century, sea travel was safer, faster, and more comfortable than land travel. To the inhabitants of eastern England, Amsterdam was much closer than London. The coastal communities of England and Scotland formed part of a North Sea nexus, linked to Holland—and, to a lesser extent, Norway, Denmark, and the Hanseatic cities—by commercial and religious ties.

The global language of liberty might easily have been Dutch rather than English but for an accident of geography: where Great Britain was an island, the Netherlands occupied a low-lying, almost indefensible plain. In the years after 1689, the Dutch Republic exhausted itself in wars against totalitarian France. By the 1720s, it was spent, its fleet deteriorating, its chief banking and trading houses relocating from Amsterdam to London.

Still, William was accustomed to notions of limited government and rule by consent. He had had to struggle for the position of stadtholder—which, while not democratic, was not exactly hereditary, either. Some historians refer to the United Provinces of the Netherlands at that time as a "crowned republic." In fact, that phrase perfectly describes Britain from 1689 to the present day—as well as the other Anglosphere monarchies.

In February 1689, Parliament drafted a Declaration of Right—
which later that year became a parliamentary statute, and so is now
known as the Bill of Rights. Its form and content, to our eyes, closely
anticipate the U.S. Declaration of Independence and Constitution—
though the authors were not looking forward, but back at the various petitions of the 1640s and, ultimately, Magna Carta.

Like the Declaration of Independence, the Bill of Rights began
by laying a series of grievances against the king. He had abused his
executive power; he had sought to tamper with parliamentary elections; he had illicitly disarmed his Protestant subjects; he had inter-
fered with the judiciary; he had prejudiced the right to trial by jury;
he had levied excessive fines and inflicted "illegal and cruel punishments" on people. It then went on to define, in the most unequivocal
terms, the sovereignty of Parliament. It declared that only Parliament might raise revenue through taxation. It rejected the idea that
an act of Parliament might be struck down other than by a subsequent act of Parliament. It protected the right of petition. It ruled
out the maintenance of a standing army in peacetime. It guaranteed the right of every Protestant subject to bear arms. It forbade the
levying of excessive bail and of "cruel and unusual punishments."
It established the principle of parliamentary privilege, and declared
that what was said in the chamber "ought not to be impeached or
questioned in any court or place out of Parliament."

These were seen as traditional freedoms, not as new ones. The
Glorious Revolution was the last and greatest of the conservative
reactions against the Stuarts. As Edmund Burke, the most eloquent
Whig of all, was to put it a century later, "The Revolution was made
to preserve our ancient indisputable laws and liberties, and that
ancient constitution of government which is our only security for
law and liberty." The Glorious Revolution was, like the Wars of the
Three Kingdoms, an Anglosphere event, touching every land where
English was spoken, albeit with local differences.

Scotland, like England, held a convention that likewise offered the throne jointly to William and Mary. The Scottish version of the Bill of Rights is called the Claim of Right. It was similar in form to its English equivalent, restating the ancient freedoms of the country and establishing the supremacy of the Scottish Parliament over the monarchy, as well as confirming the position of Presbyterianism. Unlike in England, though, where almost no one was publicly prepared to take James's part, a number of Highlanders, loyal to their supreme chief and distrustful of the English-speaking, Presbyterian majority on the convention, rose in a doomed revolt. They were called Jacobites, from the Latin form of James's name: Jacobus.

As during the 1640s, the war did not simply pit Highlander against Lowlander, but touched off the Sicilian feuds among different clans. One such feud ended in the most infamous act of treachery in Scottish history, the Massacre of Glencoe, when thirty-eight members of Clan MacDonald were murdered in their beds by rival Campbells and other government troops who had accepted the MacDonalds' hospitality.

The disturbances in Scotland were, however, nothing compared to the all-out war that racked Ireland. As their grandfathers had done in the 1640s, most Irish Catholics took the part of the Stuarts. Once again, the result was disastrous for them.

When England and Scotland declared for William, James's Irish deputy, the Catholic Earl of Tyrconnel, determined to secure the island's strong points for the Jacobite cause. He was frustrated only in Protestant Ulster, where the militia defending the walled city of Derry, led by a group of apprentice boys, withstood a siege for 105 days before being relieved by the Royal Navy. That siege has had an almost mystical importance for Ulster Protestants ever since, providing their political leaders with both their favorite slogan—"No surrender!"—and their nickname for fellow Protestants whom they regard as having betrayed the cause: "Lundy," the name of the luckless governor who had attempted to negotiate a surrender before fleeing.

Ireland was where James attempted his comeback, landing at Kinsale in March 1689 with six thousand French soldiers. There was some desultory campaigning and one major engagement at Enniskillen before, in June 1690, an impatient William arrived to take personal command of his forces, accompanied by thirty-six thousand English, Dutch, Danish, and German soldiers. William and James met at the Battle of the Boyne on July 12, 1690—the most sacred date in the Ulster Protestant calendar—and the Jacobites were routed. Once again, James fled, to the disgust of his French sponsors. ("You only have to talk to him to understand why he is here," remarked one Frenchman after meeting the exiled king at St.-Germain.)

James abandoned his Irish allies to the vengeance of the Williamites. Though clemency was offered to the foot soldiers, it was not extended to Jacobite officers, nor to Catholic landowners who had supported the old king's cause. The Jacobites fought on, hoping to secure better terms—terms that were eventually promised by William, but not subsequently ratified by the Irish Parliament. Many Irish Catholics, dispossessed of their estates, fled to Europe. Indeed, the peace treaty expressly provided for Jacobite soldiers to relocate en masse to France, and ships were provided to carry them. Some fourteen thousand men, accompanied by ten thousand women and children, chose to take service with the French Crown, theoretically in James's name, a migration known as the Flight of the Wild Geese.

The Glorious Revolution also touched the American colonies. During the Wars of the Three Kingdoms in the 1640s, Americans had participated largely as overseas volunteers, many New Englanders fighting alongside their East Anglian cousins at Naseby and Marston Moor. The skirmishing in North America itself had been localized and perfunctory. The Glorious Revolution, by contrast, was experienced as an American event, with large rebellions in Boston and New York, as well as an anti-Catholic revolt in Maryland that broke the political control of the ruling Baltimore family.

Whereas Charles I had been a distant bogeyman to the colonists, James II was a familiar foe. New York had been named after him when he was Duke of York, and he took a personal interest in the colonies while his brother was still on the throne. He particularly disliked the autonomy granted to ornery Massachusetts under its 1629 charter, and associated the Congregationalists of that province with the regicides who had killed his father. The Massachusetts charter was struck down in 1684 and, two years later, the new governor, a hard-bitten Royalist from Guernsey named Sir Edmund Andros, launched his ambitious scheme to merge the various colonies of New England, as well as New York and New Jersey, into a Royal Dominion of New England.

The historian Richard Bushman describes the new dispensation as "the realization of all Massachusetts' nightmares of oppression and avarice." Andros governed without an assembly, levied excessive fines and duties, required colonists to purchase expensive licenses to conduct their lawful business, and raised fears that he planned to establish the Anglican Church in the colony.

The 1689 risings in the British Isles were mirrored in New England and New York. The hated Dominion of New England was dissolved, and the old colonies got their separate charters again. (That of Connecticut was said to have been hidden in the bole of a white oak tree in Hartford, the famous Charter Oak.) Americans, like Englishmen, Scots, and Irish Protestants, celebrated the revolution as a restoration of their traditional liberties.

On one level, they were right: the 1689 settlement did indeed promise freedom to all English-speakers—or, at any rate, all Protestant English-speakers. Yet there were differences in how it was understood on either side of the Atlantic. Americans believed that it had vindicated their traditional rights and that, in doing so, it had tilted power back from the executive to their own legislatures. As far as most inhabitants of England were concerned, though, there was one supreme Parliament, and the Glorious Revolution had underlined its suzerainty. That difference in perception was to have huge consequences later on.

It is always possible to cavil about great moments. It is true that, from a modern perspective, the 1689 settlement was flawed. Chief among its flaws was its reconfirmation of institutionalized discrimination against Catholics.

We are now familiar with a great body of individualist Catholic social teaching. The great Spanish Jesuit leader Juan de Mariana occupies an important place in the libertarian tradition. But contemporary English-speakers had an altogether more blinkered view of Catholicism, seeing it as intrinsically authoritarian and—because Catholics recognized the supremacy of the Pope—hard to reconcile with parliamentary government.

John Locke, then attached to the Whig leader, the Earl of Shaftesbury, tied himself in knots trying to explain why religious toleration was a good idea for every other sect, but not for the Church of Rome.

Again, though, we are behaving unhistorically if we judge a past generation by the moral standards of our own. Both the English Bill of Rights itself and Locke's philosophical elaboration of the principles upon which it rested were, as the world stood at the end of the seventeenth century, extraordinary statements.

Locke's *Two Treatises of Government* were published at the time of the Glorious Revolution and expounded the model of politics that is still current in most English-speaking societies. Locke believed that all legitimate states rested upon a contract among the individuals who composed them. This contract had been made implicitly by the first of their ancestors who had agreed to live together under common rules, and had been passed down as a shared inheritance. He explained the theory from first principles:

> To understand political power and trace its origins, we must consider the state that all people are in naturally. That is a state of perfect freedom of acting and disposing of their own possessions and persons as they think fit within the bounds of the law of nature. People in this state do not have to ask permission to act or depend on the will of others to arrange matters on their behalf.

Locke's state of nature was altogether more idyllic than the primordial anarchy conjured by Hobbes. What, he asked, might have induced our early ancestors to give up their total freedom and agree to live under rules? The answer, ultimately, was the security of property, upon which all human happiness and progress depended:

> Why will he give up this empire, and subject himself to the dominion and control of any other power? To which it is obvious to answer, that though in the state of nature he hath such a right, yet the enjoyment of it is very uncertain, and constantly exposed to the invasion of others: for all being kings as much as he, every man his equal, and the greater part no strict observers of equity and justice, the enjoyment of the property he has in this state is very unsafe, very unsecure. This makes him willing to quit a condition, which, however free, is full of fears and continual dangers: and it is not without reason, that he seeks out, and is willing to join in society with others, who are already united, or have a mind to unite, for the mutual preservation of their lives, liberties and estates, which I call by the general name, property.

Locke framed his argument in abstract and universal terms, but he was well aware that it had an immediate political import, sustaining the claims of his political patron and the Whig faction. Locke's theories took concrete form in the settlement that was being negotiated at the time he was writing: one designed to safeguard the freedom and property rights of the individual.

Astonishingly, that settlement has lasted ever since. It may justly be called the most successful and enduring constitution in the world, predating its child, the U.S. Constitution, by nearly a century. What makes it especially remarkable is its authors' sincere belief that they were making no new laws, but were rather confirming the established liberties of the English-speaking peoples. All these freedoms—the common law, Magna Carta, the traditions of representative government stretching back to prehistoric times—were now given formal, constitutional force. They have it still.

Macaulay ended his *History of England* with a paean of praise to the 1689 settlement. He was writing in 1848, when Europe was being traumatized by violence and revolution, and when a Briton might feel as we feel when, warm and secure in our homes, we hear wind and rain shaking our windowpanes. Almost exactly the same amount of time separates us from Macaulay as separated him from the events he was chronicling, yet his eulogy is unsurpassed, and is worth quoting at length:

> The Declaration of Right, though it made nothing law which had not been law before, contained the germ of the law which gave religious freedom to the Dissenter, of the law which secured the independence of the judges, of the law which limited the duration of Parliaments, of the law which placed the liberty of the press under the protection of juries, of the law which prohibited the slave trade, of the law which abolished the sacramental test, of the law

which relieved the Roman Catholics from civil disabilities, of the law which reformed the representative system, of every good law which has been passed during a hundred and sixty years, of every good law which may hereafter, in the course of ages, be found necessary to promote the public weal, and to satisfy the demands of public opinion.

The highest eulogy which can be pronounced on the revolution of 1688 is this, that it was our last revolution. Several generations have now passed away since any wise and patriotic Englishman has meditated resistance to the established government. In all honest and reflecting minds there is a conviction, daily strengthened by experience, that the means of effecting every improvement which the constitution requires may be found within the constitution itself.

NOT BUILT TO ENVIOUS SHOW

Inigo Jones, the architect who was captured naked at the fall of Basing House, had designed a royal palace in Whitehall that would have outshone any in Europe. But Charles I never had the money to complete it. Like all subsequent monarchs, he was kept on too tight a financial leash by Parliament to indulge the baroque fantasies that, in the rest of Europe, were taking physical shape in marble and statuary.

Nowhere in the Anglosphere is there a kingly residence that, in scale or splendor, can rival Louis XIV's Versailles outside Paris, nor the Winter Palace in St. Petersburg, nor the Belvedere in Potsdam, nor the Herrenhausen in Hanover, nor the Buen Retiro in Madrid.

We have noted the peculiar taste for restraint in Anglosphere design, a restraint eulogized by the Country House Poets. The strength of Whiggery may, in a sense, be seen in the parsimony of Britain's royal

palaces. While Europe's princes, from Naples to St. Petersburg, were overawing their subjects with lapidary projections of their power, the British monarchy was losing property. Many of the great medieval and Tudor palaces were destroyed during the civil war, ransacked by Puritan troops or scarred by artillery. Others were sold off. As the historian Linda Colley put it, "Whereas Henry VIII had been able to hunt game, or women, or heretics out of more than twenty great houses scattered throughout England, Charles II returned in 1660 to only seven: Whitehall, St James's, Somerset House, Hampton Court, Greenwich, Windsor Castle and the Tower of London."

Although Charles wanted to restore the glory of these palaces, William III discontinued all his projects, and turned Greenwich into a hospital for disabled seamen.

To see how physically different Britain might be had the Stuarts succeeded, look at the building they *could* afford to commission from Inigo Jones—the building before which Charles I was eventually beheaded: the Banqueting House on Whitehall.

To look properly, you will need something to spread on the floor and lie upon, for the most impressive feature of the building is a ceiling that contains nine virtuoso paintings by Rubens celebrating the union of the English and Scottish crowns.

They are sumptuous works, swirling and sensual. In the main picture, England and Scotland are portrayed as fleshy women, each holding half a crown. A curly-headed lad between them is the future Charles I. Minerva, goddess of wisdom, hovers above, while below the weapons of war are consigned to a furnace.

Yet, as you sprawl on the floorboards, you find that something is bothering you. The paintings that make up the ceiling are gorgeous pieces, for which Charles I paid the almost unbelievable sum of £3,000. But the whole setup feels out of place in an English-speaking country. It is too ostentatious, too propagandist, too hierarchical in its iconography.

The more you look, the more you understand the distaste that people across the Anglosphere felt for the Stuarts. In their tastes, as well as in their politics, the monarchs seemed foreign: transalpine, ritualistic, overelaborate. It is easy to dismiss such sentiments as a kind of artistic anti-Catholicism, and they unquestionably had a sectarian component. But art is never just an expression of religious identity, and tastes are not denominational. In Rubens's native Antwerp, for example, the largely Catholic burghers built discreet townhouses rather than baroque palaces. While Rubens is never exactly restrained, the canvases he produced for his own townsmen seem sober next to the extravagant works he painted for Charles I.

At last, the paradox of the Banqueting House hits you. The ceiling was commissioned to celebrate the union of the English-speaking peoples. James VI & I liked to call himself the first Briton, and looked forward eagerly to the full amalgamation of his kingdoms. His son and grandsons fervently wished for the same. Yet their subjects, English and Scottish, regarded the entire dynasty as alien.

Whatever their individual qualities, the Stuarts were never seen as British. Their genealogy, their artistic tastes, their religious proclivities, and, above all, their political beliefs sundered them from their countrymen.

The Stuarts set out to unite the English-speaking peoples and, in the end, they succeeded, but not in the way they had intended. The Anglosphere's common political consciousness was born out of a shared hostility to the dynasty. It is significant that the parts of the British Isles from which the Stuarts drew their strongest support were non-English-speaking: the Scottish Highlands and the mass of Gaelic-speaking Ireland.

I have chronicled the anti-Stuart struggles at some length for two reasons. First, it was from these conflicts that the English-speaking peoples emerged as a single political community, recognizing a kinship that was not simply linguistic, but was based on shared values. They are values that unite the Anglosphere to this day: parliamentary supremacy, the rule of law, property rights, free trade, religious toleration, open inquiry, meritocratic appointments, representative government, control of the executive by the legislature, individual liberty.

This kinship of values was not dependent on political union. For most of their history, the English-speaking peoples have lived in connected but separate states. Nor were these values developed in England and then exported, by conquest or settlement, into a kind of greater England. In many ways, they were stronger in other parts of the English-speaking world: in Scotland, in Northern Ireland, and, above all, in North America.

The second reason that I have dwelt on the politics of the seventeenth century is that it is otherwise impossible to understand the event that created the greatest democracy on earth, the event that we commonly call the American Revolution but which ought properly to be known as the Second Anglosphere Civil War.

CHAPTER 6

The Second Anglosphere Civil War

The foundation of the British Empire was not laid in the gloomy age of ignorance and suspicion but in an epoch when the rights of mankind were better understood and more clearly defined, than at any other former period.

—George Washington, 1783

It is true that each people has a special character independent of its political interest. One might say that America gives the most perfect picture, for good or ill, of the special character of the English race. The American is the Englishman left to himself.

—Alexis de Tocqueville, 1840

HAMPDEN'S CHORD

Almost exactly the same amount of time separates us from the final defeat of the Loyalist cause at Yorktown in 1781 as separates the Yorktown combatants from the final defeat of Charles I at Preston in 1648. Our generation, especially in the United States, is generally more interested in the later of the two conflicts; but the American revolutionaries themselves were fascinated—obsessed, we might almost say—with the wars of the 1640s. The Second Anglosphere Civil War was fought over the same issues as the First; and, on both sides of the Atlantic, people overwhelmingly picked sides on the same basis that their ancestors had done.

We saw that the origins of the First Anglosphere Civil War lay in three related arguments—over tax, religion, and the location of sovereignty. The buildup to the Second Anglosphere Civil War saw precisely the same three arguments rehearsed in remarkably similar language. Both sides were conscious of the similarities. Whigs in the 1760s, borrowing the epithet that the Parliament men had thrown at Royalists, referred to their opponents as "malignants"; Tories retorted with "Oliverians."

The fighting, when it came, followed the same ethnic and religious cleavages as the war of 130 years earlier, with uncanny exactness. It did so on both sides of the Atlantic, dividing opinion in Great Britain just as in the colonies. Although the fighting led to one part of the Anglosphere declaring itself independent from the rest, it is anachronistic to think of it as a war between Americans and Britons. It was understood and described by contemporaries as a settlement by force of the Tory-Whig dispute, which by then had exhausted all attempts at peaceful resolution.

As with the First Anglosphere Civil War, we must unclutter our minds of the knowledge of what followed. We know, with hindsight, where the fighting would lead, and so can be tempted to commit Professor Butterfield's offense of studying the past with one eye upon the present. It takes a certain mental effort to picture the Anglosphere as contemporaries understood it when the quarrels began to intensify in the late 1760s.

After the Glorious Revolution, the English-speaking peoples had turned their faces to the sea. Europe, to most of them, was a source of danger, an unfree continent, teeming with tyrants, Jesuits, and exiled Jacobites. The open seas, by contrast, were a source of opportunity and commercial fortune. The people of the British Isles began to reorient from Europe to the Atlantic. Eastern cities such as Norwich declined, and wealth shifted to the great western ports of Glasgow, Liverpool, and Bristol.

English-speaking settlements and bases formed a ring around the Atlantic littoral, as well as springing up on some of the sparse islands in between. The Atlantic became almost an Anglosphere lake, its gray waves lapping against Nova Scotia, New England, Virginia, Bermuda, Jamaica, the Falkland Islands, St. Helena, Gibraltar, and various African trading posts.

Historians refer to these dominions as the First British Empire, and the American Revolution needs to be understood in the wider context of the Anglosphere as it then stood. There was no American nation before the Declaration of Independence. There were, rather, several English-speaking Atlantic colonies, stretching from subarctic Canada to the subtropical Mosquito Coast protectorate (mainly in present-day Nicaragua). When we think of what is now called the American War of Independence, we need to ignore the present-day map, which shows the United States, Canada, and the various Caribbean territories as separate states, and instead imagine the world map of the late 1760s. The differences between the colonies had to do with culture and politics rather than with divergent national identity.

We may gather the American colonies into six broad groups, some of which had a far more marked tendency to Whig militancy than others. Newfoundland, Nova Scotia, and Canada, which had been conquered from France in 1763, had little interest in radicalism. New England, by contrast, was Puritan, prickly and troublesome. New York, New Jersey, and Pennsylvania were ambivalent in their politics: their Dutch- and German-descended populations tended to be loyal to the Crown, and the Scottish Highlanders of New York became the most detested of all Loyalist troops when the fighting started. The Low Church Chesapeake gentry were broadly Whig. Florida, a garrison province seized from Spain in 1763, had little sympathy with the political agitation of either the New England Yankees or the Virginia radicals. Last were the white planters of the Caribbean, who, outnumbered by their slaves, were strongly antidemocratic.

This picture is, of course, a caricature. In every part of the Anglosphere, almost in every town, there were divided opinions. Wherever the English language was spoken, it was a medium for argument between Whigs and Tories (though they did not always use these names). Nonetheless, it is a convenient shorthand. The three main sources of revolutionary agitation were New England Congregationalists, radical tidewater planters, and the inland Ulster-Protestant settlers (more often, although inaccurately, called Scotch-Irish) spreading from Pennsylvania to the Carolinas. New England and Virginia eventually broke away to fulfill their republican aspirations; in doing so, they carried with them the less enthusiastic region between, the Middle Colonies, which had been largely deserted by the government.

The two fringes of Anglophone North America—Canada and Florida, both of which were militarized, thinly populated, and Tory in sympathy—wanted nothing to do with the patriot cause. They declined to participate in the Continental Congress, remained under the Crown, and, following the separation, offered refuge to the Loyalist refugees from the lands that had broken away.

It cannot be stressed too strongly that the American Revolution was an internal argument followed by a civil war. Only after the French became involved in 1778 did it occur to anyone to treat the conflict as one between different states. American Tories emphasized their loyalty to British institutions, above all the Crown-in-Parliament; American Whigs, by contrast, were loyal to the British values upon which the legitimacy of those institutions rested, and which they believed the king himself was violating.

When we look at the great historical panoramas painted by nineteenth-century artists, or watch the versions of the war dreamed up in Hollywood studios, we see colonists marching under the stars-and-stripes. While Betsy Ross's famous flag was certainly displayed by some Patriots, their favored banner was one that Americans have now largely forgotten: the Grand Union Flag. Known also as the Congress Flag and the Continental Colors, it had the thirteen red and white stripes as they are today, but in the top left-hand quarter, instead of stars, it showed Britain's flag, made up of the St. George's Cross for England and the St. Andrew's Cross for Scotland.

That emblem neatly demonstrates what the Patriots believed they were fighting for, namely a recognition of their rights as Britons. The Grand Union Flag was the banner that the Continental Congress met under, the banner that flew over their chamber when they approved the Declaration of Independence. It was the banner that George Washington fought beneath, that John Paul Jones hoisted on the first ship of the United States Navy. That it has been almost excised from America's collective memory tells us a great deal about how the story of the revolution was afterward edited.

Most of the places that formed part of the First British Empire are now independent. Only the tiniest—Gibraltar, the Falkland Islands, St. Helena, some Caribbean islets—remain under British jurisdiction. The British Empire had a self-dissolving quality, in the sense that the political rights and values it disseminated tended to promote local autonomy and self-reliance.

In the case of the other Atlantic colonies, separation came peaceably and by consent. Most Caribbean states secured independence in the 1960s and 1970s. Canada had internal autonomy from 1867, acquired the attributes and trappings of statehood under the 1931 Statute of Westminster, and repatriated the last elements of constitutional sovereignty from London in 1982.

The centrifugal logic that led to the eventual independence of all these territories, as well as British possessions in Africa, Asia, and the Pacific, was present, too, in eighteenth-century America. The colonists had been habituated, from the earliest settlements, to self-rule. Their soil knew neither an episcopacy nor an aristocracy, landownership was widespread, and townships, like congregations, expected to choose their own leaders. This self-reliance, however, was not the main cause of the American Revolution. Even the most radical Patriots accepted, until long after the fighting had started, that the wider British imperium of which they formed a part should be in charge of foreign policy and defense, and most also accepted that such sovereignty implied control over external trade—that is, trade between the Anglosphere and foreign territories rather than among different component parts of the Anglosphere. Britain's victories over the French had delighted the colonists, who saw them as the providential triumph of a free people over authoritarian and servile foes.

The American Revolution was made by Englishmen who, as their ancestors had done during the 1640s, asserted their rights against a monarchy that they viewed as alien and innovatory.

Alfred, Lord Tennyson, the Victorian poet laureate, came close to the truth in a slightly ponderous poem called "England and America in 1782":

> O thou that sendest out the man
>   To rule by land and sea,
> Strong mother of a Lion-line,
> Be proud of those strong sons of thine
>   Who wrench'd their rights from thee!

> What wonder if in noble heat
>   Those men thine arms withstood,
> Retaught the lesson thou hadst taught,

And in thy spirit with thee fought—
  Who sprang from English blood!

Whatever harmonies of law
  The growing world assume,
Thy work is thine—The single note
From that deep chord which Hampden smote
  Will vibrate to the doom.

Tennyson made a connection that would have been familiar to his Victorian readers, strange though it seems to us. He grasped that the American Revolution was a consummation of the English Revolution of which John Hampden, that fiery champion of parliamentary supremacy, had been the central figure.

Hampden, who led the resistance to royal absolutism in the House of Commons in the run-up to the First Anglosphere Civil War, and who was killed by Prince Rupert's soldiers at the Battle of Chalgrove Field in 1643, was a hero by any standard. His followers worshipped him. Macaulay described his emergence thus: "The nation looked round for a defender. Calmly and unostentatiously the plain Buckinghamshire Esquire placed himself at the head of his countrymen and across the path of tyranny." His opponents, too, respected his qualities. The Royalist Earl of Clarendon said of him, "The eyes of all men were fixed on him as their Patriae pater, and the pilot that must steer their vessel through the tempests and rocks that threatened it."

Hampden was an especially titanic figure to the American colonists. Towns in Maine, Maryland, and Connecticut are named after him, as is Hampden County, Massachusetts. Radical pamphleteers in the 1760s and 1770s frequently took his name as their pseudonym. There was a popular, though probably apocryphal, story of his having lived for a time in New England. American Patriots drew explicitly on Hampden's arguments against Charles I's Ship Money when formulating their "no taxation without representation" doctrines. When the fighting started, they gave his name to one of the first warships in the U.S. Navy.

This reverence for Hampden is critical to our understanding of the Second Anglosphere Civil War. Hampden had led the parliamentary resistance to the Stuarts during the linked struggles over money, religion, and power in the 1630s. Now that those three same arguments were being held again, his ideological heirs summoned his ghost to their aid.

MONEY, RELIGION, POWER

We like to remember that the American Revolution began with a taxpayers' revolt. What we often forget is that the taxpayers' revolt started in Great Britain. Eighteenth-century governments did not busy themselves with all the things that their modern successors do. There was no state funding for health care or education; policing and social security were paid for through local rates. The main financial burden on the central state was military, and an expensive foreign policy could impose huge costs on taxpayers.

Between 1756 and 1763, the Anglosphere had fought the first true world war, taking on its old rivals France and Spain. The fighting raged across Asia, Africa, America, and the West Indies, and in every theater there were stunning British victories. Canada was conquered, as were most French possessions in India, Africa, and the Caribbean. Manila and Havana were torn away from Spain. No one now doubted that the British Empire was the world's foremost power, least of all the British themselves. "We ne'er meet our foes but we wish them to stay, they ne'er meet us but they wish us away," was a typically swaggering line from the popular naval song "Hearts of Oak," composed to celebrate the triumph.

"Look around," the rising Whig politician Charles James Fox boasted to his fellow MPs. "Observe the magnificence of our metropolis, the extent of our empire, the immensity of our commerce and the opulence of our people."

But the victory had not come cheap. The national debt had risen from £72 million in 1755 to £130 million in 1764 and, for the first time, the English-speaking peoples began to understand the concept of imperial overstretch. (It was at this time that Edward Gibbon began work on his monumental *Decline and Fall of the Roman Empire*.) Quite apart from having to protect far-flung provinces against French and Spanish revanchism, the Anglosphere had for the first time annexed foreign populations whose loyalty could not be automatically assumed. Seventy thousand French-speaking Catholics had been brought under the Crown in Quebec, along with many times that number of Indian Muslims and Hindus. The costs of garrisoning this new global empire were added to the war debts.

By the end of the fighting—known in Britain as the Seven Years' War, and in the United States as the French and Indian War—the average inhabitant of Great Britain was paying twenty-five shillings a year in tax. But the average inhabitant of North America was paying only sixpence: one-fiftieth as much. Taxes in North America were negligible by contemporary standards: the total tax take, according to the historian Robert Palmer, was around one-twenty-sixth of what it was in England. It was the refusal of British taxpayers to carry this burgeoning load that led Parliament to look for ways to make the American colonies contribute financially to their own defense.

The arguments that ensued closely followed the contours of the fiscal disputes of the 1630s and 1640s. As far as we can tell, the principle mattered more than the amounts. The ministry in London was pigheaded in some of the levies it proposed, above all the Stamp Act, which required a license to be purchased for many printed materials, including playing cards, legal documents, and—an unbelievably inept addition—newspapers. Yet the strength of opposition led the ministry to back down almost immediately: the Stamp Act was repealed a year after it took effect.

Other laws that now loom large in the list of prerevolutionary grievances, such as the Sugar Act, were an early example of Laffer curve thinking. Instead of declaring high notional levies, which were then widely evaded by smugglers, Parliament sought to cut the duty, but also actually to collect it—to the horror, naturally, of the smugglers.

Indeed, the legislation that sparked the Boston Tea Party—and thus the war—was a *lowering* of the duty on imported tea.

Tempting though it always is to look for financial incentives, we are left with the conclusion that the colonists were genuinely more upset by the idea of "no taxation without representation" than by the amounts of revenue being levied from them, which remained exceptionally low by comparison with the rest of the Anglosphere, and lower still by comparison with contemporary Europe.

Why, then, did the dispute turn violent? Because, in truth, it was never really about the money. The most active opponents of the various new levies—the committees of correspondence, the secret societies, the delegates to the Stamp Act Congress—were animated by the folk memory of the anti-Stuart struggle in which their ancestors had fought. In citing Hampden's battle against Ship Money, they were consciously recalling the popular resistance that had saved the English-speaking peoples from tyranny. "What an English King has no right to demand, an English subject has a right to refuse," Hampden had argued—a phrase endlessly quoted in the 1760s and 1770s. Whigs in America, as in Britain, were convinced that the new imposts undermined their freedom and degraded the exceptionalism that was their birthright as Englishmen.

We have seen that, under James I and Charles I, the arguments over revenue were as bitter as they were because they were tinged with sectarian differences. The same was true in the 1760s and 1770s. In both cases, the Crown found itself in the awkward position of holding both arguments simultaneously, often with the same opponents.

The religious dimension of the American Revolution has been neglected by most historians. No one much likes to dwell on the bellicose Protestantism in the colonies. The distinguished professor of British and American history J. C. D. Clarke concluded that "the virulence and power of popular American anti-Catholicism is the suppressed theme of colonial history."

It is hard to disagree. We have already noted the furious reaction to the recognition of the Catholic Church in Quebec. In 1774, delegates to the First Continental Congress raged against what they saw as a sickening betrayal:

And by another Act the Dominion of Canada is to be so extended . . . that by their numbers daily swelling with Catholic emigrants from Europe, and by their devotion to Administration, so friendly to their religion, they might become formidable to us, and on occasion, be fit instruments in the hands of power, to reduce the ancient free Protestant Colonies to the same state of slavery with themselves.

Again, we need to remember that popular anti-Catholicism was political, not doctrinal. In the colonies, as in Great Britain, it was widely supposed that Catholics were not reliably patriotic, that their ultimate loyalty was to foreign powers. John Jay, one of the Founding Fathers, who went on to become the first Chief Justice of the United States, used precisely the same reasoning as Locke to argue that his home state of New York should extend full toleration to every sect except the professors of the religion of the Church of Rome, who ought not to hold lands in, or be admitted to a participation of the civil rights enjoyed by the members of this State, until such a time as the said professors shall appear in the supreme court of this State, and there most solemnly swear, that they verily believe in their consciences, that no pope, priest or foreign authority on earth, hath power to absolve the subjects of this State from their allegiance to the same.

John Adams, the second president, wondered, "Can a free government possibly exist with the Roman Catholic religion?"

Thomas Jefferson, the third, believed that Catholicism was inseparable from political authoritarianism: "In every country and in every age, the priest has been hostile to liberty. He is always in alliance with the despot, abetting his abuses in return for protection to his own."

These views, though widespread, were not universal. James Madison, the fourth president, seems to have been free of religious prejudice, and George Washington was disdainful of sectarianism of any kind. Indeed, it is in no small measure thanks to the political stature of its first president that the United States was born not only as a decentralized republic but also as a state without legal religious discrimination.

Washington was consciously holding up an example for his more hotheaded countrymen. He had come across plenty of religious bigotry, and well understood the worldview of those who saw the rise and fall of nations as signs of God's favor or disapprobation, and military triumphs as providential, a sign that the Reformation had been divinely ordained.

We, in our age, find it much harder to enter into that mindset. Nor can we easily understand why contemporary English-speaking Catholics and High Churchmen were so nervous about letting individuals read the Bible and make their own decisions about forms of worship. In fact, that belief was entirely understandable in the context of its time.

The very belligerence with which some Protestant sects pursued their convictions served to convince many Catholics and Anglicans that disseminating the scriptures without context or instruction was unsettling. The book of Revelation is strong stuff, promising a "New Jerusalem" that will arise when the "Whore of Babylon" has been overthrown. Some contemporaries saw these passages not as theology but as a political manifesto, and equated the "Whore of Babylon" with whomever they happened to dislike. Without context and interpretation, parts of the New Testament did indeed incite millenarian violence.

There were good reasons, it seemed to many, why the people had for centuries not been allowed to swallow unlimited drafts of these texts but had instead had them rationed out and strained through the cloth of sacerdotal Latin. Look what had happened, they remarked, once vernacular Bibles became widespread.

The seventeenth-century poet Samuel Butler had spoken of how the contemporary Puritan would strive "to prove his doctrine orthodox / By apostolic blows and knocks."

The same was true in Massachusetts during the following century. And here, too, the "Whore of Babylon" was taken by some militant Protestants to mean not just Rome but any Christian sect whose practices were deemed too ritualistic.

Just as seventeenth-century Puritans had suspected High Church Anglicans of having Romish leanings, so their descendants carried that conviction into North America. The Anglican Church was established in some of the colonies, a cause of constant grievance to the Congregationalists, Presbyterians, and other Dissenters who saw it as a bulwark of authoritarian government.

For these men, ideological and usually lineal heirs of the Roundheads and Covenanters, the war was a spiritual one, a crusade against idolatry and superstition. George III's Church of England was, in their eyes, hopelessly corrupt, repressing intellectual freedom just as the Tories longed to repress political freedom. The Quebec Act had confirmed what Nonconformists had always half suspected, namely that High Church Anglicans, with their altar rails and bishops and stately communions, were to all intents and purposes allies of Rome.

The rows over the Sugar and Stamp Acts were, for these psalm-singing men, the battlefield that God happened to have ordained for them, just as taxation happened to be the issue that the Pharisees had used to tempt Jesus. It was the religious dimension that made the fiscal disputes so intractable. The great historian of religion in America, William Warren Sweet, put it well: "Religious strife between the Church of England and the Dissenters furnished the mountain of combustible material for the great conflagration, while the dispute over stamp, tea and other taxes and regulations acted merely as the matches of ignition."

As in the 1630s and 1640s, the mixture was highly flammable. And, as then, the fiscal and religious disputes had a way of spilling over into an argument about the location of sovereignty.

In both cases, historians, knowing where that argument led, tend to emphasize the constitutional rows over the sectarian differences that sustained them. Yet, in the minds of contemporaries, the fiscal, denominational, and democratic arguments were different aspects of a single issue. Protestantism, low taxes, property rights, and parliamentary self-government were, in Whig minds, fused into a libertarian alloy; an alloy that could not now be melted back down into its separate elements.

Legislative independence was assumed by most Americans to be a birthright their forebears had carried to Plymouth Rock. But it was also seen as the surest way to secure their religious and fiscal rights. Which is why, as the arguments became more venomous, the colonists moved beyond the questions of tax and the creation of

American bishops—questions that, by 1775, had largely been settled in their favor—and focused instead on the question of parliamentary sovereignty.

Looking back in 1800, Madison, the future president, came up with as neat a summary of the constitutional objectives of the Patriots as you could ask: "The fundamental principle of the Revolution was that the colonies were co-ordinate members with each other and Great Britain of an empire united by a common sovereign, and that legislative power was maintained to be as complete in each American parliament as in the British parliament."

Stated thus, it seems a remarkably moderate proposition, and many in Britain saw it as such. Whigs in the House of Commons moved to accommodate what they saw as the colonists' just demands. In 1775, William Pitt the Elder proposed to repeal every piece of legislation that the American Patriots had found objectionable, beginning with the Sugar Act, and to recognize the Continental Congress as, in effect, an American parliament, coequal with Britain's.

Even at this late hour, the acceptance of such a scheme would probably have secured the continued unity of the Anglosphere. The American colonies would have drifted nonviolently to eventual independence in the way that Canada and Australia did. The Kingdom of America might very well today form part of the Commonwealth.

But the House of Lords, Tory and authoritarian, had no intention of compromising with what it saw as unlawful rebellion. Pitt's proposals were voted down, 61 to 32. The last real chance for a political settlement was lost.

The balance of opinion in Britain's hereditary chamber should not blind us to the sympathy that existed for the colonists' grievances, not just at popular level, but within the political nation. The American cause united the greatest parliamentarians of the time: Pitt himself, the dashing radical leader Charles James Fox, and the intellectual grandfather of the conservative Anglosphere tradition,

Edmund Burke. America may, indeed, have been the *only* issue that brought these three titans together.

We need to remember that the quarrel was still seen by all sides as a family row. We must not think, anachronistically, of British radical sympathy with the American Patriots as being support for a foreign ally. Whigs formed a single faction within a single polity, and felt equally threatened by a ministry that seemed bent on returning to Stuart Toryism.

"I rejoice that America has resisted," Pitt had proclaimed when, a few years earlier, he had torn into the Stamp Act. "Three million people so dead to all feelings of liberty as voluntarily to submit to be slaves would have been fit instruments to make slaves of the rest [of us]."

Burke, one of the greatest orators ever to have graced Parliament, made powerful speeches on behalf of the colonists, whom he unquestioningly took for fellow countrymen. Indeed, Americans were, for him, more British than those who had remained in the mother island, for they had exaggerated the peculiar concept of liberty that was the distinguishing feature of English-speaking peoples: "The colonists emigrated from you when this part of your character was most predominant; and they took this bias and direction the moment they parted from your hands. They are therefore not only devoted to liberty, but to liberty according to English ideas, and on English principles."

Burke was in no doubt that, in pressing their rights, the American radicals were asserting rather than denying their English heritage. It was, he told MPs, a peculiarity of England that its constitutional development had turned on the question of taxation, and its liberties had grown out of the struggle to ensure that only the people's elected representatives might raise revenue: "The colonies draw from you, as with their life-blood, these ideas and principles. Their love of liberty, as with you, fixed and attached on this specific point of taxing."

And as they had exaggerated that aspect of their national identity, so they had exaggerated the religion that was thought to be inextricably linked to political freedom: "The people are Protestants; and of that kind which is the most adverse to all implicit submission of mind and opinion. This is a persuasion not only favorable to liberty, but built upon it."

Burke was convinced that the only way to settle the unrest was to confirm the rights of the colonists as full British citizens: "Let us get an American revenue as we have got an American Empire. English privileges have made it all that it is; English privileges alone will make it all it can be."

Burke's views were, as far as we can tell, popular in the country at large. But they had less support in the House of Commons, whose members were elected on a limited franchise and, in some cases, saw professions of exaggerated loyalty to the king as a route to preferment and emoluments. More than a century after the Restoration of Charles II, ancient cleavages could still be glimpsed on the leather benches: one study of the minority of MPs who opposed the repeal of the Stamp Act in 1766 found a roll call of Cavalier surnames: Bagot, Curzon, Grosvenor, Harley.

The government was prepared to make concessions on the *level* of the new taxes, but not on the *principle* of its right to levy them. In the eyes of George III and his ministers, the way to deal with dissent was through firm measures. They had, they believed, done more than enough to meet the religious and fiscal concerns of the colonists. What they would not do was surrender the sovereignty of the British Parliament, which, in their eyes, was the supreme council of the Anglosphere. It might license subordinate local chambers in the colonies. But there was a difference between devolution and federalism.

It was true, ministers conceded, that people in Boston had no representatives in the House of Commons. But then, neither had most people in, say, Birmingham, England. Both, argued ministers, were "virtually represented," in the sense that they had champions and sympathizers there. Patriots retorted that Birmingham, too, should have actual representation. Pitt described the notion of "virtual representation" as "the most contemptible idea that ever entered into the head of a man; it does not deserve serious refutation."

But Lord North and his government remained obdurate. Like unpopular regimes in every age, ministers convinced themselves that they were dealing only with a handful of troublemakers and that, if these were dissuaded by a show of force, the bulk of the population would be loyal. That calculation turned out to be spectacularly wrong. Just as had happened in the 1640s, events had already taken on a force of their own. The argument was no longer about taxation or religion, but about power. The decision to send General Gage to disarm the Massachusetts militia was the precise equivalent of Charles I's attempt to control the English militia, and was seen as such. All over New England, men swarmed to join armed bands thrown together to resist the regulars. The Second Anglosphere Civil War had begun.

THE SECOND COUSINS' WAR

Once the fighting had started, the Crown's cause in New England was lost. The only way to hold down a population in armed revolt would have been through repressive measures that were unthinkable in the English-speaking world, even among the most extreme supporters of the regime. The war that followed was, in reality, fought to determine which other colonies would follow New England, and on what terms the partition would be made.

Far from being prepared to terrorize the American population, most inhabitants in Great Britain sympathized with the Patriot rather than the Loyalist cause. The government had enormous difficulty finding recruits, especially in England. Like the Stuart Kings, George III was fatally obliged to look for soldiers in Scotland, Ireland, and—the issue that swung moderate opinion in the Americas against him—Europe.

Responding to Pitt's proposed redress of colonial grievances in 1775, the former lord chancellor, the Earl of Camden, had issued a strikingly wise and accurate prophecy:

> To conquer a great continent of 1800 miles, containing three millions of people, all indissolubly united on the great Whig bottom of liberty and justice, seems an undertaking not to be rashly engaged in. It is obvious, my Lords, that you cannot furnish armies or treasure competent to the mighty purpose of subduing America, but whether France and Spain will be tame, inactive spectators of your efforts and distractions is well worthy the considerations of your Lordships.

Wars have a way of hardening opinions on both sides. Like the Roundheads of the 1640s, American Patriots found that they had started down a road that led inexorably to republicanism. They could not treat the king as a military enemy one day and invite him to resume his rule over them the next. In London, too, the mood quickly turned, especially after the New Englanders invaded Canada in the fall of 1775. Having passed the Sugar and Stamp Acts in the sincere belief that they would make possible a lower land tax in Great Britain, the country gentlemen in the House of Commons now loyally approved a rise in the land tax to four shillings to pay for the war effort.

Where the First Anglosphere Civil War had united the English-speaking peoples in a single state, the Second ended in a rupture. One consequence is that later generations of historians have tended to cover only their side of it. We have already seen how, in the United

States, the story of Paul Revere's ride has been doctored to make it seem as if the fighting were between Americans and Britons—a concept that no one at the time would have recognized. The very title "War of Independence" is misleading, for it implies that there was an extant American nation being ruled by a different nation in the sense that, say, the Congo was ruled by Belgium.

Most American historians have focused overwhelmingly on events on their side of the Atlantic, giving some consideration to the Loyalists (the most hard-line of whom conveniently departed after the war), but little to the pro-American tendency in Great Britain. Most British historians have made the equivalent error. That is to say, while they have explored the motives of the radicals in Great Britain, they have tended to treat their support for American Whigs as sympathy with an overseas cause, much as later generations of British radicals were to campaign for Spanish Republicans or anti-apartheid South Africans.

One writer who approached the conflict in terms that contemporaries would have understood—namely as a civil war with ancient roots within the Anglosphere—was the former Reagan administration official Kevin Phillips. In 1995, disgusted by the tawdriness of Clinton-era Washington, he retreated to his Connecticut home to write a book about whether, had "the British" been a bit tougher during the Saratoga campaign in 1777, the American Revolution might have been forestalled. But the more he studied the period, the more ambitious his project became. As he explored the reasons that British generals had been so reluctant to prosecute the war more vigorously, he began to see that the origins of the American Revolution were in fact to be found in the British Isles. Having tramped across the battlefields of Saratoga and Yorktown, he soon found himself doing the same at Naseby and Marston Moor (Britain, unlike the United States, is shamefully neglectful of these sites:

where American battlefields are beautifully tended and covered with memorials, British ones are left as turnip fields and are often bisected by new roads).

Being a political strategist rather than a professional historian, Phillips stumbled upon something that eluded most specialists, namely the essential continuity between the two Anglosphere civil wars. In both cases, Puritan militiamen with names like Isaiah and Obadiah were battling against what they saw as a corrupt, tyrannical, crypto-Catholic monarchy. In both cases, the battle touched every part of the Anglosphere: England, Scotland, Ireland, and North America. And, in both cases, these places tended to divide along the same geographical and denominational lines.

Indeed, Phillips took the argument one stage further. For him, the American Civil War was also a continuation of the earlier two Anglosphere civil wars. Once again, he saw accustomed battle lines taking shape, as the New England Yankees, who had led the fighting against Charles I and George III, invoked their "God of War" in a new crusade. Once again, he saw the Episcopalian landowners of the South using the arguments that Royalists had used, namely that they were defending an orderly, settled, natural way of life against agitators and enthusiasts. Once again, he saw factions in Britain ranging themselves familiarly, with Nonconformists and evangelicals backing the Union while Tories looked kindly on a confederacy whose leading men had toyed with the idea of making Victoria their queen in return for British recognition.

The result of all these researches was the seminal 1999 book *The Cousins' Wars: Religion, Politics, and the Triumph of Anglo-America*. Its central thesis—that the English Civil War, the American Revolution, and the American Civil War were three spasms in one ongoing conflict—sounds implausible when baldly stated. But Phillips had done his homework, tracing the loyalties of church groups and even individual families from one war to the next, and finding extraordinary political continuities. The mark of a convincing new thesis is that, though it may jar at first, it afterward seems obvious. Once the American Revolution is understood as a civil war, much falls into place.

Phillips looked in detail at who had chosen which side in the Americas, and found that, in most cases, their sympathies were inherited from the Wars of the Three Kingdoms. Strongest for the Patriot cause were the New England Yankees; the Low Church tidewater gentry of Maryland, Virginia, and North Carolina; and the Ulster Protestants. In George Washington's darkest moment, leading the remnants of the Continental Army away from the debacles of Germantown and Brandywine, he despairingly wrote: "If defeated everywhere else, I will take my last stand for liberty among the Scotch-Irish of my native Virginia."

Strongest for the Crown were the High Anglicans, German Lutherans, Irish Catholics, and Scottish Highlanders (Jefferson had to be made to remove a specific attack on Scotland from the Declaration of Independence).

It goes without saying that these are broad generalizations. Almost every county was divided. Some families fell out. Many people simply wanted to be left alone. John Adams's assessment that "we were about one third Tories and one third timid and one third true blue" is not far off the mark. The best guess of historians nowadays is that around 20 percent of the active white population of the Thirteen Colonies was Loyalist, around 40 percent Patriot, and 40 percent neutral. It was George III's wavering between petulance and weakness that eventually pushed most of the "timid" into the Whig camp.

As often happens in civil wars, the national conflict overlaid and absorbed local vendettas that had little to do with the issues at stake. If one group picked a side, its local rivals would often pick the other.

Many Native Americans, for example, fought for the Crown, seeing it as an ally against land-hungry settlers. But their decision pushed some tribes, for reasons of inherited enmity, onto the Patriot side. Many slaves fought with the Loyalists hoping for emancipation, which drew some slave owners into a more radical position than they might otherwise have chosen. As one historian of American loyalism put it, "In every feuding neighborhood they [Loyalists] were one of the two local parties; for irrelevant disputes were generally not abandoned at the onset of war: instead they quickly took on, almost at random, the larger enmities of Whig and Tory."

One especially fateful decision was the Crown's refusal to attempt a large-scale occupation of the colonies after the first shots were fired at Lexington and Concord. The ministry hung back, hoping for reconciliation, but in practice ensuring that armed revolt spread well beyond Massachusetts. Phillips was in no doubt about the consequence:

> From May 1775, when the events at Lexington and Concord became known, to November 1777, as the word of victory at Saratoga spread, the majority of the thirteen colonies were unoccupied for most of those 30 critical months. No significant troops were present to interrupt rebel consolidation of political power. Hardly any were on hand to inhibit collection of local tax revenues, control of local militia, and procurement of food supplies, weapons, and munitions. Historically, this has not been the way to quell a revolution.

Indeed. But George's ministry had a problem. There was little appetite in Great Britain for a repressive war against fellow countrymen whose grievances struck most people as just.

Several senior military officers who had served against the French in North America—Lord Amherst, Sir Henry Conway, and others—flatly refused to take up arms against the colonists. Those who did agree carried out their commissions to the letter, but without enthusiasm. Generals Gage, Carleton, and Howe made clear that they were unhappy with the war, and in consequence were heartily disliked by American Tories. Even Generals Burgoyne and Clinton had no stomach for the kind of coercive force that would have been required to bring the rebels to heel. One gets the impression that these were men doing their duty rather than fighting to win.

Their distaste for the task was widely shared in the British Isles, which divided much as they had during the 1640s. Of course, these things are not easy to gauge scientifically. The British Parliament was elected with a very restricted franchise, and the views of the unrepresented classes cannot be precisely measured. Still, we can get an idea from the circulation of the newspapers that favored one side or the other; from petitions to Parliament, either for coercion or for conciliation; and from the attitudes of that minority of MPs who, under the Byzantine electoral system that pertained before the rationalization of 1832, represented more populous constituencies.

What we find is telling. Public opinion in Great Britain seems to have been remarkably similar to that in the colonies, with perhaps 25 to 30 percent of the population broadly Tory in inclination, and the rest Whig with varying degrees of enthusiasm. The main reason that the American legislatures were so much more radical than the House of Commons turns out to be a rather banal one: land was more evenly distributed in America, and the colonial assemblies were elected by a far higher proportion of the male population. They were therefore more representative of public opinion as a whole.

Even more fascinating is how the British Isles divided geographically, for here we see an almost exact replication of the battle lines of the First Anglosphere Civil War. The areas that were strongest for the American rebels were those that had resisted the Stuarts most fiercely: the Scottish Lowlands (especially the southwest); London and its surrounding counties; the Puritan flatlands of Cromwell's old Eastern Association; the Nonconformist cloth towns; and Ulster, whose martial inhabitants were soon forming militias and drilling in mimicry of their Pennsylvanian cousins.

The parts of England that, by contrast, were keenest on coercion, at least if we judge by the evidence of their petitions to Parliament, were the old Stuart redoubts: the West Midlands and, above all, Lancashire. With recruiting sergeants in England struggling to find men for the American campaign, Lord North was forced to turn to the old Jacobite heartlands of Ireland and the Scottish Highlands.

James II's heirs had never abandoned their claim to the throne, and the first half of the eighteenth century had seen constant Jacobite plots and invasion scares, as well as full-scale risings in 1715 and 1745. The second of these ended in the obliteration of the Jacobite cause when Charles Stuart, Bonnie Prince Charlie, the grandson of James II and son of James Stuart whose birth had prompted the warming-pan story, was routed at Culloden Moor. The distance of that defeat gives it a patina of Romantic imagery: the glint of claymores, the mist rising from the heather, the dashing escape of the tartan-clad prince. But most contemporary English-speakers reacted with dizzy relief. The old threat to their liberty and property—or, more prosaically, to their Protestantism and their commercial success—had finally been lifted.

With Jacobitism finished, most Highlanders swung behind the Hanoverian dynasty with the same stolid fidelity they had offered the Stuarts. They fought with terrifying ferocity against their former French allies in the Seven Years' War as, indeed, they have in almost every major British engagement since. As William Pitt the Elder put it in 1766:

I sought for merit wherever it could be found. It is my boast that I was the first Minister who looked for it and found it in the Mountains of the North. I called it forth, and drew into your service a

hardy and intrepid race of men; men who left by your jealousy
became a prey to the artifices of your enemies, and had gone nigh
to have overturned the State in the War before last [the 1745 rising].
These men in the last War were brought to combat on your side;
they served with fidelity as they fought with valor, and conquered
for you in every quarter of the world.

The American unrest was an opportunity for the sons of the
men beaten at Culloden to prove their faith again, while at the same
time striking a blow against the descendants of the Covenanters,
their ancient foes. Highlanders made up a majority of the regular forces that saw action in the American campaign, forming no
fewer than ten regiments. They were complemented by volunteer
battalions formed by their émigré kinsmen. Large numbers of de-
mobilized Highlanders had purchased land in the colonies after
the Seven Years' War, some along the Hudson Valley, others in the
Carolinas. These men now rushed to answer the king's call, a few
wearing the Jacobite white cockade in their bonnets. They formed
several auxiliary Tory regiments: the North Carolina Highlanders,
the Royal North British Volunteers, the Highland Company of the
Queen's Rangers. This last is now a Canadian regiment, the Queen's
York Rangers, for many of these loyal men chose exile in the north
rather than life in a republic.

Ireland, too, was largely for the Crown. Once more, the evidence
is necessarily patchy. Because Catholics were excluded from political life, the MPs and town corporations reflected only the views
of the Protestant minority—who were, for the most part, fiercely
Whig. Again, though, when we look at petitions, at the declarations of Catholic priests and, not least, at the number of military
volunteers, there is no denying the enthusiasm for the Royalist cause
in Ireland—though, naturally, generations of Irish Americans
have done their best to deny it, focusing on the exceptions and, on

occasion, blurring the distinction between Scotch-Irish and Irish. Still, studies by Owen Dudley Edwards and Conor Cruise O'Brien are conclusive. Irish Catholics were overwhelmingly Loyalist and, indeed, their loyalty at last won them relief from some of the penal legislation that had been in place since the Flight of the Wild Geese. Laws passed during the American war lifted many of the civil disabilities from which Irish Catholics suffered and restored their right to bear arms and serve as soldiers.

Phillips's book contained no new primary research. All of its facts were drawn from published works of history, some of them dating from the early twentieth century. Yet, in bringing the data together open-mindedly, Phillips gave a sense of perspective to the Anglosphere Civil Wars—the Cousins' Wars—that had eluded many academic historians.

Like the First Anglosphere Civil War, the Second was fought in a gentlemanly fashion—not just by the standard of its time, but by the standard of civil wars generally. The routine roundups and massacres that accompanied, say, the Spanish Civil War in the 1930s were unknown, and the occasional lapses by one side or the other were newsworthy precisely because they were occasional. Some Tories had their property confiscated; some unpopular officials were tarred and feathered. But, as in the English Civil War, the machinery of local government, including the sheriff's offices, functioned throughout the war.

Casualties were, in the context of their age, almost unbelievably light. According to the U.S. Defense Department, there were 4,435 fatalities and 6,188 other casualties on the Whig side. Tory losses were even slighter. When we think of the tens of thousands who were dying in Britain's wars against France at the time, Yorktown was a skirmish.

Though Hollywood would have us believe otherwise, the military authorities on both sides did their best to behave chivalrously. When the governor of Quebec, Sir Guy Carleton, was asked why he treated Whig prisoners so well, he replied, "Since we have tried in vain to make them acknowledge us as brothers, let us at least send them away disposed to regard us as first cousins"—which is, more or less, what happened.

In all wars, of course, there is sequestration and looting, but such atrocities as there were did not happen at the hands of regular forces on either side. Rather, they were the result of local internecine feuds that often had little to do with the national issue.

Nathanael Greene, Washington's most trusted general, was horrified by the vicious ambuscades and cattle raids he found in South Carolina: "The animosity between the Whigs and Tories of this state renders their situation truly deplorable. There is not a day passes but there are more or less who fall a sacrifice to their savage dispositions. The Whigs seem determined to extirpate the Tories and the Tories the Whigs."

In the end, of course, the Whigs prevailed. And, unlike in the 1640s, they were now in a position to remove the more dangerous of their enemies. While 80 percent of the Loyalist population remained in the republic, the most doctrinally committed Tories left, either by choice or from social pressure. Those leaving the southern states generally relocated to Florida, the West Indies, and the Bahamas, often taking their slaves with them. Most of those fleeing the Middle Colonies and New England headed north, some to Quebec, more to Nova Scotia, where a new province was eventually carved out to accommodate them: New Brunswick. A few went as far as the British Isles.

The Harvard historian Maya Jasanoff estimates that a total of 60,000 Tories, including 10,000 black Loyalists, emigrated. Of these, 33,000 settled in Nova Scotia and New Brunswick; 6,600 in Quebec; 5,000 in Florida; and 13,000 (including 5,000 free blacks) in Great Britain. A number of the black Tories ended up founding a settlement in Sierra Leone.

At the same time, Whig and radical emigration from Great Britain to what was now the United States of America began to increase following independence. Perhaps five million British migrants crossed the Atlantic in the century and a half after the Declaration of Independence, and while the migration was overwhelmingly economic, the United States always exerted a stronger pull for British radicals than did alternative destinations.

The Tory-Whig division, though its balance had varied from place to place, had existed throughout the Anglosphere. After 1776, however, the Whigs won more or less total political control of one part of the English-speaking world. What followed was the most perfect consummation of English Whig philosophy in the form of the U.S. Constitution and, most especially, the Bill of Rights.

All states develop according to the DNA that was fixed at the moment of their conception. The United States of America was founded on a series of premises: that concentrated power corrupts; that jurisdiction should be dispersed; that decision makers should be accountable; that taxes should not be raised, nor laws passed, save by elected representatives; that the executive should be answerable to the legislature.

The men who met in the old courthouse at Philadelphia were determined to prevent a repetition of the abuses through which they had lived. In consequence, they came up with the most successful constitution on earth: one that, to this day, has served to keep the government under control and to aggrandize the citizen. The peculiarities of the American governmental model—states' rights, the direct election of almost every public official, an accountable judiciary, primaries, ballot initiatives, balanced budget rules, term limits—all are a working out of the Jeffersonian ideal of the maximum devolution of power. If the Second Anglosphere Civil War was the genotype, they are the phenotype.

Yet again, we need to remind ourselves that the Founders saw themselves as conservatives, not innovators. In their own eyes, all they were doing was guaranteeing the liberties they had always assumed to be their heritage as Englishmen. Far from creating new rights, they were reasserting rights that they traced back through the Glorious Revolution, through the First Anglosphere Civil War, through Simon de Montfort's campaigns, through even the Great Charter itself to the folkright of Anglo-Saxon freedoms.

I have tried to show that their version of history was nowhere near as fanciful as is often claimed. The exceptionalism of pre-Norman England was real enough. The English-speaking peoples were indeed set apart by their political structures. In a sense, though, what mattered most was that these things were *believed* to be true.

The histories most widely read in the colonies—Nathaniel Bacon's *Historical Discourse of the Uniformity of the Government of England*; Henry Care's *English Liberties*; Lord Kames's *British Antiquities*—all told the same story: in 1066, a free people had lost their liberties to a Continental invader, and their subsequent history had been a struggle for the restoration of those liberties. Even at the time of independence, there were several Americans who were aware of having non-English ancestry, yet they cheerfully bought into a self-consciously Anglo-Saxon political identity.

That identity was now exalted. The seventeenth-century campaign to "throw off the Norman Yoke" had finally been vindicated. William the Conqueror's lieutenants and their Tory descendants had been banished.

On both sides of the Atlantic, many regretted that these outcomes had required a rupture. One of the most wistful and beautiful lines that Jefferson had put in the Declaration of Independence was eventually excised from the final version: "We might have been a great and free people together."

By then, though, events had moved on. As Lord Camden had foreseen, France and Spain were not "tame, inactive spectators" during the fighting. The involvement of these ancient foes had the same effect on moderate opinion in Great Britain that George III's infamous decision to use German mercenaries had had on moderate opinion in the colonies.

France formally declared war on Great Britain in 1778, and Spain in 1779, and soon engagements were being fought from the Caribbean to Gibraltar, from India to Central America. In this new world war, North America became a sideshow. Even George III slow-wittedly came round to the idea that his sovereignty in the colonies was over. "It is a joke to think of keeping Pennsylvania or New England," he declared. "They are lost."

That large body of transatlantic opinion, which, as late as 1775, had hoped for some form of Anglo-American federation, was quickly dispersed before the volleys of foreign muskets. The sundering of the Anglosphere became an accepted fact—although a further war had to be fought in 1812 before British Tories accepted emotionally what they had accepted legally, namely that the United States was a wholly independent state rather than a sort of tolerated protectorate; that it had the same naval rights as any other power; and that British subjects who emigrated there became foreign citizens, and were no longer subject to conscription.

The political unity of the English-speaking peoples had been short-lived. After its Second Civil War, the Anglosphere reverted to being a legal, cultural, and linguistic entity rather than a single state.

Its partition had happy consequences for both sides. In Britain, the downfall of Lord North's rotten ministry was followed by urgent administrative reform. The power and prestige of the monarchy declined, that of Parliament expanded, and there was a focus on meritocracy and efficiency. Pitt the Elder's son, William Pitt the Younger, became prime minister in 1783 at the age of just twenty-four. He was to lead the government, with one intermission, until his early death in 1806, and was as restless, brave, and brilliant an occupant as 10 Downing Street has known. He restored the national finances, prepared the ground for the abolition of slavery in 1807, and defeated a resurgent, revolutionary France.

The independence of the United States turned out to be a great boon for the Anglosphere's military cause. Instead of having to divert significant troops and resources to North America, Britain was able to concentrate on other fronts, knowing that the Americans could be relied upon to press their own claims vigorously against France and Spain—which nations, indeed, they soon ousted altogether from their continent.

The split accelerated the shift from mercantilism and monopolies to free trade. Anglo-American commerce had recovered to prewar levels by 1785, doubled by 1792, and has been exceptionally strong ever since.

The separation also gave the Anglosphere a boost in a more direct sense, leading to the colonization of two of its core members. The Loyalist exodus to Canada ensured that that great expanse become part of the Anglosphere. Had forty thousand English-speakers not trudged into the snowy north, the entire territory might have remained essentially Gallic in language and culture. And the loss of Georgia, whither Britain had been in the habit of transporting its able-bodied criminals, created the need for a new penal destination. Transportation to Australia, whose more habitable eastern part had been claimed by Britain in 1770, began five years after the recognition of American independence.

Above all, the nature of the war, and the arguments that had preceded it, produced the greatest constitution ever drafted, designed to prevent the concentration of power and written in full awareness of man's fallen nature. Where most national constitutions come and go every few decades (or, in South America, every couple of years), the U.S. Constitution has served the purpose for which it was intended for more than two centuries—that is to say, it has ensured that the

government is constrained and the citizen is free; that jurisdiction is dispersed; that decisions are taken as closely as possible to the people they affect; that power is balanced.

That constitution was not just an American achievement. It was, as its authors were keen to stress, the ultimate expression and vindication of the creed of the English-speaking peoples. The ideals of the rule of law, representative government, and personal liberty, ideals that had their genesis in those ancient forest meetings described by Tacitus, had found their fullest and highest expression.

| CHAPTER 7 |

Anglobalization

I have often before now been convinced that a democracy is incapable of empire.

—CLEON, QUOTED BY THUCYDIDES, FIFTH CENTURY BC

The English colonies—and this is one of the chief reasons for their prosperity—have always enjoyed more internal freedom and more political independence than those of other nations

—ALEXIS DE TOCQUEVILLE, 1835

FROM ANGLES TO ANGLOSPHERE

When friends from overseas ask me about the prospects of Scotland leaving the United Kingdom—at the time of writing, the country is preparing for a referendum in September 2014, seven hundred years after Robert the Bruce's victory over Edward II at the Battle of Bannockburn—their assumption is usually that independence is popular with the Scots but unpopular with the English. In fact, opinion polls tell us that the opposite is the case. Scottish voters oppose separation by a solid two-to-one majority. In England, the breakdown is much closer to half and half.

While opinion polls didn't exist in the seventeenth or eighteenth centuries, the evidence we do have suggests that—other than occasional blips, such as that which followed Scotland's oil strike in the early 1970s—the union of the two ancient realms was always more popular north of the border than south. The persistence of the idea that the Union was some sort of English takeover tells us more about the way our age imagines hierarchies of power and victimhood than about what actually happened.

The story of the amalgamation of Scotland and England is worth retelling, for it refutes the idea that the Anglosphere is somehow just an amplification of England.

If we define nationality, as it was traditionally defined, by language and ethnicity, the borders within the British Isles make no sense. We might arguably draw a frontier around the Gaeltacht: the fragmentary Irish-speaking peatlands and islands to which, in the nineteenth century, earnest middle-class nationalists would bring their children, sometimes by coracle, to teach them the language of their ancestors. It would certainly be feasible to draw one around the breathtaking Welsh mountains where the rich tongue of the ancient Britons is still heard. We might even get away with a border around the parts of the Hebrides and the scant adjoining Highland villages where Scottish Gaelic clings on.

The people who lived in these remote, beautiful, sparse pockets, preserving the pre-Saxon talk, were generally regarded by their English-speaking neighbors as rustic and backward. Not until the Romantic revival of the nineteenth century did the antique tongues become fashionable, attracting the attention of folklorists and lexicographers.

The British Isles now constitute an almost uniformly Anglophone territory, albeit one in which some older languages happen to be

taught in schools. The Celtic languages are spoken at home by less than one percent of the population. Irish children learn Irish much as English children used to learn Latin and Greek: as an accomplishment valuable in itself rather than as a skill with practical, everyday utility.

It is true that there are immense varieties within the vernacular English of the British Isles, some dialects of which are almost impenetrable to the outsider. I am addicted to the traditional patois of Lowland Scotland, variously known as Braid Scots, Doric Scots, and Lallans. It is an idiom aptly suited to the temper of my mother's people, and two of my favorite courtroom stories involve stony Scottish judges. One, impressed by the unexpected eloquence of the defendant, told him: "Ye're a cliver chiel, but ye'll be nane the waur o' a guid hangin" (You're a smart fellow, but you'll be no worse for a good hanging). Another, trying a man for sedition, was irked when the defense claimed that Jesus himself had had to answer similar charges: "Aye, an' muckle guid it did Him: He was hangit!" (Yes, and much good it did Him: He was hanged!)

Over the past twenty years, various public sector bodies in Scotland have treated Braid Scots, along with Scottish Gaelic, as a minority language, and it is a delight to find this wry, terse form of English written down. But it is just that: a form of English. When we see it on the page, in even its most demotic form, it closely resembles the speech of, say, County Durham in England, or County Donegal in Ireland.

The people of the British Isles are united by culture as well as speech. They watch the same television programs, shop at the same chains, eat the same diet, abuse alcohol in the same way, read the same newspapers, dress in similar fashion. Yet these mores are not the basis of their national identity. The United Kingdom has been readier than most European countries to extend nationality—not just paper citizenship, but a genuine sense of common belonging—to peoples of different ethnic, linguistic, and religious origins; a process

that has been taken even further in the newer Anglosphere countries, and furthest of all in the United States. Why did this extension
happen so naturally? Because, while Englishness and Scottishness
could certainly be defined, like other nationalities, with reference to
place and race, Britishness could not.

The identity that was projected onto overseas territories was thus,
from the beginning, a political rather than an ethnic one. For all that
people down the centuries have been prone to say "England" when
they mean "Britain," no one ever spoke of the "English Empire."
Even before the values of the Anglosphere were carried overseas,
they had transcended any single nation. This gave the Anglosphere
an advantage over every rival state, blurring the distinction between
mother country and colony. If nationality was chiefly defined by a
set of political values, then by implication, anyone who signed up to
those values could belong. Nowadays, that notion might not sound
terribly shocking, but in the age of the great European empires, it
was utterly revolutionary.

HATH NOT GOD FIRST UNITED
THESE KINGDOMS?

When James VI of Scotland became James I of England in 1603, he
wanted to rule one new dominion, not two ancient ones. In his first
speech to the House of Commons as sovereign, he declared:

> Hath not God first united these Two Kingdoms both in Language,
> Religion, and Similitude of Manners? Yea, hath he not made us all
> in One Island, compassed with One Sea, and of itself by Nature so
> indivisible, as almost those that were Borderers themselves on the
> late Borders, cannot distinguish, nor know or discern their own
> Limits?

Many of his Scottish subjects agreed. Though cross-border raid-ing had been a way of life for centuries, there was a far more imper-meable cultural frontier between the Highlands and the Lowlands than between the Lowlands and England.

Although we nowadays think of kilts and plaid and bagpipes as Scottish national emblems, this iconography was in reality the invention of one man: the nineteenth-century writer Walter Scott. Until his time, most Lowland Scots—which is to say, most Scots—regarded such dress as the badge of a thief.

The real border was the linguistic and geographical one that cut through Scotland. The word *Sassenach* (Saxon), now an all-purpose derogatory Scottish term for the English, was, before the twenti-eth century, used by Highlanders to describe both Lowlanders and Englishmen, between whom they drew no distinction. The Low-landers, for their part, felt far more in common with the English than with the clansmen from whom they were sundered by custom and speech.

Many Scots saw, in the union with England, huge advantages: wealth, a large home market, overseas colonies, and, with a Scottish king on the throne, plum jobs in the state machine.

The English saw things rather differently. As landless Scottish lairds and gentry swarmed south with their monarch, snapping up sinecures and titles, they felt almost as if they were being invaded.

James wanted to merge his kingdoms formally, and to call him-self "King of Great Britain." He was proud of his book learning and, at the earliest opportunity, reminded his English subjects of the days of their heptarchy: "Do we not yet remember, that this King-dom was divided into Seven little Kingdoms, besides Wales? And is it not now the stronger by their Union? And hath not the Union of Wales to England added a greater Strength thereto?"

English MPs were unconvinced and denied the king the title he craved. Nevertheless, though England and Scotland remained

formally separate realms, their political and cultural approximation accelerated.

We saw in the last chapter that the great civil and religious disputes of the seventeenth century cut laterally across both nations.
They did not pit Englishmen against Scotsmen, but radicals in
both countries against Royalists in both (as well as in Ireland and
America).

A sense of common British nationality long preceded the formal
merger of the two countries in 1707. That's not to say that the union
was universally popular; far from it. There is always an attachment
to existing institutions: what Milton Friedman was later to call "the
tyranny of the status quo." It is again worth noting that this resistance was stronger in England than in Scotland.

The immediate catalyst for the pooling of the national legislatures was an ill-fated attempt to found a Scottish colony at Darien
in Panama in 1698. Many Scots felt that they had been left out as
Spain, Portugal, France, England, and the Netherlands established
bases and trading posts across the oceans. An eccentric scheme was
concocted to plant a Scottish settlement, to be known as "Caledo-
nia," on the isthmus. The colony would supposedly control the neck
of land, and carry freight by mule train from ocean to ocean for
vast fees. Why the Spanish should tolerate this challenge to their
jurisdiction was never properly explained. Nor did anyone ask why
the English should risk antagonizing Spain over a colony that, if
successful, would benefit only Scotland.

The enterprise was a wretched failure. The settlers, fair-skinned
and accustomed to the cold, were not well suited to jungle conditions. Most died of tropical diseases, dysentery, or starvation. The
few who survived surrendered when a small Spanish force was sent
to close down their base. Of the 2,500 colonists who had set out from
Leith, huddled belowdecks so that the English should not get word
of what they were up to, fewer than four hundred returned.

By then, though, large chunks of the Scottish upper and middle classes had invested their life savings in the enterprise. The failure of the Darien colony was not simply a national humiliation; it was, for many influential Scottish families, a financial catastrophe. From both patriotic and personal motives, they concluded that their best hope lay in piggybacking on their larger neighbor, and turned to England for help in stabilizing their currency and wiping out their debt.

As in 1603, England was the more hesitant partner. Why, Englishmen wanted to know, should they bail out a government less thrifty than their own? Why was the value of the Scottish pound any of their business? And what advantage was there for them in allowing Scots access to their colonial markets?

Queen Anne, however, was as determined as any of her Stuart ancestors to unite her two kingdoms. The English administration appointed an overwhelmingly unionist negotiating team, and terms were eventually agreed: Scotland would retain its church structure, legal institutions, and educational system, but would join a full commercial and political union with England. The Edinburgh parliament would dissolve itself and Scottish MPs would instead sit at Westminster.

A legislature does not vote lightly for its own disbandment, and getting the two separate Acts of Union through the English and Scottish parliaments required much cajolery and a fair amount of outright bribery. State sinecures, hereditary titles, and several thousand pounds in hard cash were dispensed on both sides of the border.

Oddly, the bribery was remembered and resented in Scotland, but not in England. Robert Burns wrote a bitter poem about the "parcel of rogues" who were "bought and sold for English gold." Yet a referendum in England, had such a thing been possible at the time, would almost certainly have rejected the Union by a large majority. The idea that England had annexed its less populous neighbor, so widespread today, would have struck contemporaries on both sides of the River Tweed as preposterous.

Well into the second half of the eighteenth century, there was a low-level, smoldering resentment in England against Scottish arrivistes. The archetype of the Scotsman on the make makes its first appearance in English theatrical productions of the 1760s. Radical English Whigs—curiously, from our modern perspective—portrayed all Scots as authoritarian Tories and secret Jacobites. English cartoons of the era invariably show Scotsmen in kilts and tam o'shanters—dress that had in fact been proscribed in Scotland except, tellingly, in the Highland regiments of the British Army.

As Linda Colley showed in her great work *Britons*, English anxieties were by no means irrational. Scottish universities were in those days far better than their English rivals at educating doctors and engineers, and these professionals naturally sought employment south of the border. Military and colonial posts were disproportionately filled by Scotsmen. In the decade after 1775, 47 percent of the men appointed to serve as clerks and administrators in Bengal were Scots, as were 60 percent of the men allowed to reside there as traders. And Whig prejudices against the autocratic tendencies supposedly fostered by Highland culture were not wholly baseless: many of the most inflexible Royalist governors in North America had indeed come from Scottish Jacobite backgrounds.

The Union ushered in a cultural renaissance in Scotland. David Hume dominated in the field of philosophy, William Robertson in history, Adam Smith in economics. The Anglosphere as a whole was lit up by the dazzle of the Scottish Enlightenment. The streets and squares of Edinburgh's New Town, that marvel of classical architecture dating from the 1760s, proclaim in their names contemporaries' attachment to the Union and its Hanoverian dynasty: Prince's Street, George Street, Queen Street, Hanover Street.

The promise of Rubens's Banqueting House paintings, which had shown the arms and artifacts of war being consigned to a furnace, was fulfilled. At this distance in time, it is easy to lose sight of what was, for contemporaries, easily the strongest argument for the Union, namely that it would put an end to centuries of cross-border bickering. And so it did. Freed from their quarrels, England and Scotland turned outward, much to their mutual gain.

In time, the English, too, came to see the benefits. In creating unprecedented wealth for themselves, Scots generated wealth for the Anglosphere as a whole. Today's fifty-pound bill commemorates the collaboration between James Watt, the brilliant Glasgow engineer, and his English partner Matthew Boulton. Watt's inventions enriched Scotland; but they also enriched England's new industrial city of Birmingham, where he had settled. If the fifty-pound note were torn in two, neither half would be worth twenty-five.

When the English and Scottish populations came to see each other as fellow countrymen, their ancient rivalry mellowed into the good-natured bantering that has lasted to this day. The relationship between the two nations became rather like that between the English Dr. Johnson and his Scottish biographer, James Boswell: teasing, competitive, occasionally grouchy, but fundamentally affectionate. It is a relationship beautifully captured in the words of the Highland soldier who, observing the rout on the beaches of Dunkirk in 1940, told his sergeant: "If the English give in, too, this could be a long war."

We have seen before that people often recognize a common identity in the face of something truly foreign. So it was in the eighteenth century. Britons defined themselves, in the popular culture of the time, as everything that the Continental monarchies in general, and the French in particular, were not. Where the Europeans were supposed to be effete, autocratic, obsequious, superstitious, and ignorant, the English-speaking peoples were frank, manly, independent, freethinking, and enterprising.

It is hugely significant that this self-image was exaggerated in America, Scotland, Wales, and the Protestant parts of Ireland. After the final defeat of Jacobitism, the peoples of these regions liked to think of themselves as especially sturdy, especially undeferential, especially devoted to Whig and liberal principles. If Britain as a whole had a Protestant culture, they were more intensely Protestant—Congregationalist in New England; Presbyterian in Scotland, Ulster, and Pennsylvania; Nonconformist in Wales—and so, by implication, more intensely British.

Why, then, is it nowadays so often seen the other way around? Why do we think of England having imposed itself on the peripheries rather than the peripheries having disproportionately defined the orientation of the new state?

In part, it is a question of sheer size: in 1750, there were five Englishmen for every Scot; today, after generations of migration, there are eight. By the time of the Acts of Union, London, for reasons that historians still haven't adequately explained, had become a colossus, far larger than any contemporary European city. The metropolis attracted aristocrats and skilled professionals from the fringes, and for the first time, there was large-scale intermarriage among the great English, Scottish, and Anglo-Irish families, creating a self-consciously British ruling class. From the perspective of many Englishmen, it looked like a takeover by pushy Celts. But from the perspective of tenant farmers in Scotland, Wales, or Ireland, it seemed as if their local landowners had gone to England and become English.

This dual imbalance—of size and of social status—is a clue to one of the more striking aspects of national identity in the Anglosphere, namely the rejection of patriotism by the left. In most of Europe, the national cause was a popular and democratic one. But in the Anglosphere it became entangled with questions of supporting the underdog and establishing hierarchies of victimhood. It is rarely easy to portray either Britain or the United States as an underdog. It is especially hard to think of England this way. Scottish and Welsh nationalists have therefore recast their causes as the struggles of small countries to retain their identity. The idea that the union mainly benefited its smaller members, though widespread in the eighteenth and nineteenth centuries, doesn't fit our present prejudices.

Consider the common error of saying "England" when what is really meant is "the United Kingdom." This usage is widespread in Europe, and not uncommon even in the Anglosphere. It was very frequent in the United Kingdom itself until the end of the twentieth century, when the creation of devolved assemblies in Edinburgh, Cardiff, and Belfast made English people much more conscious of the niceties.

What is interesting, though, is the way people react. When I was a boy, the use of "England" to mean all four United Kingdom nations was a source of constant irritation to the inhabitants of the other three, even those of a strongly Unionist bent. Pointing out the solecism would usually prompt the speaker casually to say that he was terribly sorry, old boy, quite right, no harm intended—which, of course, annoyed his hearers even more.

Yet, between the mid-eighteenth and the early twentieth centuries, the same elision was regularly made by Scottish, Welsh, and Irish writers who, though well aware of the distinction, regarded quibbling about it as pedantic. Pluck almost any public figure at random—Scots such as Macaulay or John Buchan, Irishmen such as Lord Palmerston or Oscar Wilde—and we find them describing themselves quite unselfconsciously as English.

Admiral Nelson's message to his fleet before the Battle of Trafalgar in 1805 is perhaps the most famous naval signal in history: "England expects that every man will do his duty." (His subordinate Admiral Collingwood's response on seeing it is less well known, but far more in tune with the national character: "I wish Nelson would stop signaling. We know well enough what to do.") No one at the time took it as an affront to the other constituent nations in the United Kingdom. As a Scottish MP put it in the House of Commons, "We commonly when speaking of British subjects call them English, be they English, Scotch or Irish; he therefore, I hope, will never be offended with the word English being applied in future to express any of His Majesty's subjects, or suppose it can be meant as an allusion to any particular part of the United Kingdom."

When did people start taking offense? Mainly in the late twentieth century, as the empire declined and the British brand found itself increasingly derided and traduced. Again, hierarchies of power and victimhood came into play. Scotland and Wales were especially adversely affected by the degeneration of heavy industry, and Englishness became associated with remote elites: Conservative politicians, Anglican landowners, London bureaucrats.

By the 1990s, the peoples of the four constituent parts of the United Kingdom were reviving their older patriotisms. A chance draw between the English and Scottish soccer teams in the 1996 European championship, which that year happened to be hosted by England, prompted the unprecedented sight of a stadium filled with red-and-white St. George's crosses—a flag that, until then, had been largely neglected. Before 1996, English fans had carried Union flags to sporting events, and the English flag was a rarity; since, it has been the other way around.

Which brings us back to the opinion polls. There are, inevitably, grievances on both sides of the border. Some Scots resent the fact that, while they overwhelmingly back left-of-center parties, they regularly get Conservative governments returned by English votes. Some English people complain, conversely, that they send their taxes north of the border and get nothing in return but socialist MPs.

These, though, are hardly the kinds of great national questions that accompanied the secessionist struggles of, say, Chechnya or Tibet or Kosovo. It is impossible to imagine a shot being fired in anger over the question of whether Scotland should remain in the United Kingdom. And the reason it is impossible is that the ethnic and religious distinctions that defined the Chechens or the Tibetans or the Kosovar Albanians don't exist. The Union is more than an amplified alliance between kindred peoples. There is also such a thing as British identity.

It's true that, compared to its constituent identities, Britishness is a rather legalistic concept. It cannot compete with Englishness, Scottishness, or Welshness in poetry or song. When people in the United Kingdom remember evocative landscapes, when they feel the pulse of lyrical place-names, when they become emotional over sport, it is generally these older patriotisms that animate them. Britishness, by contrast, is a political and constitutional construct, based primarily on shared political values and institutions.

Yet it is no less real for that. The convergence of the English and Scottish people hastened after the Acts of Union, and an undeniable pride in the new state infused its citizens. Linda Colley comprehensively showed that, as the eighteenth century wore on, this pride, being based on the notion that Britain was a providential nation, was widespread among all classes and both sexes, and was especially pronounced in Scotland and Wales.

It is worth stressing how exceptional it was to have a form of patriotism that was based on something other than blood and soil. The British saw themselves as being set apart by unique institutions: a sovereign parliament, the common law, secure property rights, an independent judiciary, armed forces that were subordinate to the civil authorities, Protestantism, and, above all, personal freedom.

It is often said that the United States is unique in being a propositional nation. Yet this characteristic did not develop in a vacuum. Long before the Declaration of Independence, Britons on both sides of the Atlantic had taken pride in the idea that theirs was a creedal rather than an ethnic identity. The unification of the English and

Scottish realms, which had for so long been not only separate but also antagonistic, more or less obliged the inhabitants of both kingdoms to define their new identity in political rather than racial terms.

It was as a newly united, multiethnic state that the British built their empire. The Americas had, in a manner of speaking, been carried by the English with them into the Union. The later colonies were unambiguously British. Herein lies the significance of the fact that no one ever described the British Empire as English.

The administration of the empire, and the commerce it sustained, were disproportionately Scottish. The armed forces that garrisoned it were not only disproportionately Scottish but even more disproportionately Irish. At the beginning of the nineteenth century, England constituted around 60 percent of the population of the British Isles, Ireland 30 percent, and Scotland 10 percent. Yet the largest concentration of British troops, namely the Bengal army, was 34 percent English, 48 percent Irish, and 18 percent Scottish.

Scots, in particular, saw opportunities across the oceans, and seized them eagerly. Fifteen percent of the British settlers in Australia were Scottish, 21 percent of settlers in Canada, and 23 percent in New Zealand. (For the Irish, the figures were 27 percent, 21 percent, and 21 percent.)

Having expressed their national identity in political terms, the colonizers could hardly fail to impart a measure of this identity to the colonized. If Britishness was defined by equality before the law, representative government, property rights, and the rest, then as they became imperial subjects, the Jamaicans, Maltese, and Malayans must surely acquire a measure of them.

This was the key to the British Empire's ultimately self-dissolving quality. Being British was not a question of birth or ancestry, as was being, say, German or Polish. Britishness meant a series of political entitlements. Anyone might become British, just as anyone might become American, by taking on these values. And since these values included parliamentary representation, there were only two possibilities: an imperial parliament, overwhelmingly dominated by the numerically preponderant Indians; or the eventual maturity of the colonies into independent states.

Although the empire had been acquired in a jumbled, unplanned way—"in a fit of absence of mind," as the Victorian essayist J. R. Seeley put it—British policy makers during the nineteenth century came to see their role as being based on stewardship. Once the colonies reached the requisite level of political development, the aim should be to oversee their development into sovereign allies.

The radical MP J. A. Roebuck said in 1849, "Every colony ought by us to be looked upon as a country destined, at some period of its existence, to govern itself." His sentiments were echoed by the Tory Arthur Mills in 1856: "To ripen those communities to the earliest possible maturity—social, political and commercial—to qualify them, by all the appliances within the reach of a parent State, for present self-government, and eventual independence, is now the universally admitted object and aim of our Colonial policy."

It is understandable that at least some of the former colonies now like to think of the independence process as a struggle led from the periphery rather than a privilege granted from the core. And, in some cases, there was indeed a struggle. There were wars fought in Cyprus and Palestine and an especially brutal one in Kenya.

These wars, though, were the exceptions. In most colonies, independence came peacefully and by agreement, and the new administrations gladly retained their links to the Commonwealth and Anglosphere. There was only one place that rejected such links: a territory seen by British policy makers not as a future sovereign state, but as an integral part of the United Kingdom. The independence of Ireland did not come peacefully and by agreement.

It came about after a bloody rising and was followed by two civil wars: a short and intense one to the south of the border, a slow and intermittent one to the north.

The Republic of Ireland was the Anglosphere's most reluctant member. Its early political leaders aimed to sever their links with the rest of the English-speaking world, to build up an autarkic economy, and, indeed, to revive the Irish language so that Ireland should literally be beyond the Anglosphere.

How last-century all that now sounds. As W. B. Yeats put it, "All changed, changed utterly."

THE FIRST COLONY

In 2011, the Queen and the Duke of Edinburgh toured the Republic of Ireland—the first such visit by a reigning British monarch since George V's in 1911, when all Ireland was still in the United Kingdom. The monarch visited Croke Park, where, in 1920, the Royal Irish Constabulary had opened fire on the crowd watching a Gaelic football match; fourteen civilians were killed. She laid a wreath in memory of the people who had died during those wretched years, and rounded off the visit with a walkabout in Cork, the spiritual capital of Irish republicanism.

The visit was a stunning success and marked the moment when the United Kingdom and Ireland, so long sundered by politics, recognized that, despite everything, they were allies. Though far-left groups had sulkily boycotted the queen's tour, almost all the Irish media hailed it as a success, and were especially delighted when the monarch began her official speech in Irish.

Not that quarrels between the two governments had ever stood in the way of good relations between their citizens. In every year since the partition in 1921, more Irish nationals have lived under British jurisdiction in Great Britain than in Northern Ireland. Following their separation, the two states immediately gave each other reciprocal social security entitlements, university access, even voting rights. There isn't a town in England or Scotland that doesn't have Irish connections, and almost no Britons regard the Irish as foreign. The two countries, indeed, had everything in common *except* their politics.

Once again, we must guard against the study of the past with one eye upon the present. Because twenty-six Irish counties broke away from the United Kingdom in 1921, it is tempting to see every event, from the resistance to Henry II's invasion onward, as a buildup to that rupture.

In fact, the English conquest of Ireland—or, more properly, the Norman conquest of Ireland—was part of a general European trend. We saw earlier that Northmen, exploiting an advantage in military technology, were seizing land all over Europe. What happened in Ireland was not so very different to what happened from Scotland to Cyprus. In 1169, a deposed Irish king, Dermot Mac-Murrough of Leinster, invited some Norman barons from England and Wales to help him regain his kingdom. These men, though garnished with the grand aristocratic titles of their era, were still essentially mercenaries. They agreed to terms, crossed the water, defeated Dermot's enemies, and were rewarded with lands and castles. Henry II, alarmed lest one of his barons should acquire the whole island and become a rival, landed with his own army in 1171 and accepted the fealty of the Irish lords and bishops. For the next eight centuries, England and Ireland were under the same crown.

This was not, however, by any stretch of the imagination, the beginning of "English oppression" of republican folklore. The English and the Irish, in common with many European peoples at that time, were alike oppressed by Norman barons—those "magnificent blond beasts," as Friedrich Nietzsche was later luridly to put it, who could be found "at the root of every aristocratic line."

Ireland had never been united before the Normans: there was no independent Ireland to hark back to. Tensions only really began after the Reformation, which affected Great Britain profoundly but barely touched Ireland. As we have seen, England fought chronic wars against Spain and France, from Queen Elizabeth's reign until the final defeat of Napoleon at Waterloo. Throughout much of this era, Ireland was seen as a weak point: a place where invading Catholic armies might land and secure popular support. During the reigns of Elizabeth and James I, Protestant settlers—some from England, many more from Scotland—were planted in Ulster and along Ireland's east coast as a defense against such an invasion.

From then on, there were sectarian quarrels, rebellions, and brutal counterinsurgencies. But, again, we should guard against anachronism. Irish nationalism in the sixteenth, seventeenth, and eighteenth centuries did not generally take the form of seeking separation from the Crown. As we have seen, Irish Catholics were overwhelmingly for Charles I during the First Anglosphere Civil War, and overwhelmingly for the Loyalist cause in the Second.

The Irish problem, from London's perspective, was easily enough stated. It was recognized that the majority of the Irish population did not enjoy full democratic rights. But granting such rights—in other words, extending the franchise to Catholics—might lead to the victimization of the Protestant minority on the island.

Pitt the Younger came up with what might have been a workable solution: he would extend the right to vote to Catholics while merging the British and Irish parliaments. That way, Irish Catholics would get civil equality within a larger electoral unit. They would enjoy full political freedom, but not be in a position to expropriate Protestant landowners, against whom they had long-standing grudges.

We'll never know whether Pitt's scheme would have settled the Irish problem, for it was frustrated in the House of Lords. The Act of Union went through, dissolving the Irish Parliament and granting Ireland representation at Westminster, but Catholic emancipation was held up for another three decades.

It was the beginning of a recurrent pattern. Britain would eventually move to address Irish grievances, but always too late. W. E. Gladstone, the great Liberal prime minister and supporter of Irish Home Rule, complained that the trouble with the Irish question was that, whenever you found the answer, they changed the question. He was right, in the sense that British policy consistently lagged behind Irish opinion. At first the agitation had focused on the removal of the remaining civil disabilities suffered by Catholics. By the time the civil rights issue was addressed, the question had moved on to land reform. Legislation was eventually introduced here, too, but now the campaign was about Home Rule. Home Rule was belatedly granted in 1912, despite the fierce opposition of Ulster Protestants who, consciously mimicking their seventeenth-century Scottish ancestors, had signed the Ulster Covenant, promising to resist any moves to Irish autonomy. As with Pitt's proposal for emancipation-with-union, we'll never know whether the 1912 legislation would have settled the issue, for the First World War intervened before it could be implemented, and events took on a momentum of their own.

What we do know is that, right up until 1916, Irish nationalist demands were focused on a restoration of the pre-1800 settlement— that is, an Irish parliament under the Crown. Republicanism was a fringe position. John Redmond, who led the Irish party in the run-up to the 1916 Easter Rising, wanted Ireland and England to remain on the closest possible terms. When the First World War broke out, he toured Ireland, urging men to join up and fight for the alliance of English-speaking peoples—which, in overwhelming numbers, they did.

Although republicans now backdate the popular support for their cause, the records tell a different story. In April 1900, Queen Victoria visited Dublin, wishing to recognize the exceptional heroism of her Irish soldiers in the South African war. Here is a contemporary account of her reception: "When we got into Dublin the mass of people wedged together in the street and every window, even on the roofs, was quite remarkable. Although I have seen many visits of this kind, nothing has ever approached the enthusiasm and even frenzy displayed by the people of Dublin."

There were some dissenting voices, but not many. As Arthur Griffith's *United Irishman* newspaper noted glumly two weeks after the visit, "We have learnt a strange and bitter lesson; let it not be lost upon us. There is much to be done to absolve the land from the treachery of the last few weeks."

Such absolution, in Griffith's eyes, was to come sixteen years later. The Easter Rising was catalyzed by the First World War in more ways than one. Republican leaders feared that, once large numbers of Irish veterans returned having served under the Union flag, the dream of complete separation would be over. Although it is rarely mentioned, more southern Irish Catholics died in British uniform *on the first day* of the Somme offensive than participated in the Rising.

The British authorities saw the Easter Rising as an act of unspeakable treachery at a time when British armies, including hundreds of thousands of Irish soldiers, were engaged in a life-and-death struggle in France. Nonetheless, their response was disgracefully and, in the circumstances, idiotically heavy-handed. The rebels were shot in batches over several weeks. Some of them were no more than teenage boys. Public opinion throughout Ireland turned suddenly and implacably.

It might be argued that the Easter Rising, and the eventual recognition of the Free State, flowed from the past refusal of the Tories to agree to Home Rule. Perhaps so. But there was nothing inevitable about the sundering of the United Kingdom, let alone about the terror and bloodshed that followed, when Queen Victoria made her stately procession through Dublin. For all the talk of eight hundred years of oppression, the events that precipitated the partition, and loosed the blood-dimmed tide, came in a late and jumbled rush.

"Was it needless death after all?" asked Yeats. Probably. For the better part of a century, quarrels in Ireland degraded and impoverished all the peoples of the British Isles. Only in the past decade have relations been restored to what ought, by any normal consideration, to be their natural state.

For the truth is that, in the eyes of most British people, though Ireland may be a separate country, it is not a foreign one. The Irish talk as we talk, dress as we dress, eat as we eat (and, tragically, drink as we drink). We watch the same television programs, follow the same football teams, shop at the same chains. We share that half-humorous, half-cynical mode of conversation that sets us apart even from other Anglosphere nations.

In fact, Britain and Ireland have been joined by pretty much everything that lies beyond the remit of government: history and geography, habit and outlook, commerce and settlement, blood and speech. It's significant that you usually hear Irish words in the context of some state office or government function: while our people have carried on their custom of intermarriage and intermixture, the two governments for a long time remained stubbornly apart.

Perhaps it was inevitable, at least at first. The early Irish leaders were, if not always anti-British, at least determined to flaunt their separateness by distancing themselves from whatever Britain was doing (though, during the Second World War, many Irish citizens felt differently, and rushed to enlist in the British Army, winning 780 decorations, including seven Victoria Crosses). I don't think it's going too far to say that Irish politicians were initially attracted to the EU partly because the Brits disliked it—though it wasn't long before, like all politicians, they also acquired a personal stake in the system.

Yet, since the turn of the century, the old antagonisms have been

wiped away. The euro crisis pushed the two kindred nations together, and the queen's visit sealed their alliance.

An opinion poll in 2010 showed that 43 percent *of Sinn Féin
voters* wanted to swap the euro for the pound. The brutal way in
which the other euro-zone governments turned on Ireland during
that crisis—accusing it of being a low-tax parasite, and demanding that it raise its corporation tax—was widely contrasted in Irish
newspapers to Britain's unhesitating offer of financial assistance.

Almost every day brings some new rapprochement. The Irishmen who had fought for the Anglosphere in the two world wars were
finally honored and given official recognition. At the beginning of
2012, Ireland went further, rehabilitating the five thousand men who
had deserted the Irish armed forces in order to fight for Britain in
the Second World War. The gesture was pregnant with significance:
Enda Kenny's government was coming as close as it could to repudiating Éamon de Valera's policy of neutrality. For the five thousand
deserters had, under any definition, broken the law. If they were now
to be honored, the implication was that the law itself had been uncon-
scionable, and that they should have been able to fight against fascism
as part of an Irish army joined to the Allied cause.

The five thousand deserters were unusual only in that they
already belonged to another army. More than seventy thousand
southern Irish Catholics flocked to the recruiting offices of Northern Ireland, and many more to those of England, during the Second
World War.

My late father, whose family roots were Ulster Catholic, saw action
alongside some of them, serving in Italy with a religiously mixed regiment, the North Irish Horse. I asked him once whether he could
remember any sectarian differences among his fellow soldiers. The
only thing he could think of was that, on one occasion, some of the
men had good-naturedly taken turns singing rebel and loyalist songs.
Faced with a properly foreign enemy, their differences had dissolved.

The Irish Free State frustrated the various schemes for imperial confederation in the 1920s. While many Australians, Canadians, New Zealanders, and South Africans toyed with the idea of some kind of continuing political union, southern Ireland was determined to go in the other direction. The partition had been followed by a brutal civil war south of the border between, in effect, those who favored autonomy under the Crown and those who demanded total independence. The latter, initially defeated, eventually established their supremacy, and they intended to ram home their victory.

The British authorities would not contract a deal with the Dominions that excluded Ireland, and so the various schemes for confederation faded. Which is why it is so significant that an Irish prime minister can now stand alongside his British counterpart at 10 Downing Street and, without making any fuss, describe the United Kingdom as his country's most important ally. The days when you could get a cheer from any Irish political rally by attacking whatever the British foreign policy of the day was are over. The country no longer has anything to prove. Ireland has rejoined the Anglosphere.

ALL OUR POMP OF YESTERDAY

As a political coup de théâtre, it surpassed even the glories of the Mughal emperors. In January 1877, four hundred Indian princes, rajas, begums, and kunwars converged on Delhi with their retinues. Each strove to outshine his peers. There were magnificent jeweled turbans and dazzling robes and sashes. There were black-bearded bodyguards with round ornamental shields and scimitars. There were golden howdahs, with silken tasseled sunshades, swaying on the backs of solemn elephants.

There to meet the native lords were fifteen thousand British soldiers, stiff in in their red and gold uniforms, alongside ranks of plumed and medaled imperial officers. There, too, were the native troops on whom the empire depended: tall Indian cavalrymen with green and blue and ochre tunics and bright pennons fluttering from their lances.

The Delhi durbar had been summoned to celebrate the proclamation of Queen Victoria as Empress of India and, to any observer, it would have seemed the epitome of imperial power and stability. "India has until now been a vast heap of stones," the Maharajah of Indore told the viceroy, a Romantic Tory poet named Lord Lytton. "Now the house is built and, from the roof to the basement each stone of it is in the right place."

Rudyard Kipling wrote a short story in which wild tribesmen from beyond the frontier marvel at the discipline of a review of Indian troops. How, a tribal chief asks a native officer, was the miracle achieved whereby even the animals turned and wheeled as one? The native officer proudly replies:

> "They obey, as the men do. Mule, horse, elephant, or bullock, he obeys his driver, and the driver his sergeant, and the sergeant his lieutenant, and the lieutenant his captain, and the captain his major, and the major his colonel, and the colonel his brigadier commanding three regiments, and the brigadier the general, who obeys the Viceroy, who is the servant of the Empress. Thus it is done."
>
> "Would it were so in Afghanistan!" said the chief, "for there we obey only our own wills."
>
> "And for that reason," said the native officer, twirling his mustache, "your Amir whom you do not obey must come here and take orders from our Viceroy."

For sheer spectacle, the Delhi durbar was hard to beat, and it was repeated on an even grander scale a quarter of a century later, under the viceroyalty of the equally Romantic Tory Lord Curzon, to mark the accession of Edward VII as king-emperor. But, magnificent as they were, these pageants could hardly have been less apt as emblems of British rule. The dominion they symbolized— grand, monarchical, military—might have done just as well for any empire. Those who believed that Britain was different, that its legitimacy rested on its libertarian values, found them more than a little uncomfortable.

We have seen that, from at least the seventeenth century, there had been two broad political tendencies in the English-speaking world. One emphasized individualism, representative democracy, and Anglosphere exceptionalism and, for the sake of abbreviation, might be labeled Whig—though that title was not always used by its partisans. The other emphasized stability, hierarchy, and order, and might—again, sometimes anachronistically—be labeled Tory.

When the English-speaking peoples built their homes across the oceans, they carried these two tendencies with them. In America, they hardened into the Patriot and Loyalist factions of the Second Anglosphere Civil War. But they were carried, too, to territories with advanced indigenous civilizations, chief among which was India.

In British-administered lands with large native populations, the main battleground between Whiggery and Toryism was the political status of the indigenous peoples. Whigs wanted to get to the point where local people would be able to sustain a Westminster-style representative democracy, and saw education as the key. Tories believed it was hopelessly naive to impose British values on alien cultures and sought instead to use traditional power structures, co-opting native chiefs and princes to their cause.

The Delhi durbars were the supreme visual manifestation of this latter approach, which Niall Ferguson calls "Toryentalism." Yet it was the Whig approach, ultimately, that made the Anglosphere what it is today: a network of independent states united, not by political structures, but by a common approach to liberty and property. That approach had transformed the settler colonies, one by one, into self-governing parliamentary states. It was bound to reach India in the end.

TRUE PATRIOTS WE

In 1796, Sydney opened its first theater in what is now Bligh Street. A troupe made up almost wholly of ex-convicts staged Edward Young's tragedy *The Revenge*. The venture was not a success: the audience was, perhaps unsurprisingly, rife with pickpockets, and the theater closed soon afterward. The performance is now remembered for a couplet from a prologue read out before the play to mark the occasion:

> *True patriots we: for be it understood,*
> *We left our country for our country's good.*

The line is popularly attributed to George Barrington, an Irish actor and cutpurse who, after repeated arrests (one of his tricks was to work dressed as a clergyman), was eventually sentenced to seven years' transportation in 1790. In fact, the words were almost certainly someone else's: the style is not Barrington's. But the fact that they were written at all in a penal colony is remarkable, and subsequent generations of Australians have adopted them fondly.

How did a nation founded as a prison camp become one of the wealthiest and freest places on earth? How did a colony conceived as a spillover for Britain's undesirables become an indispensable and unhesitating British ally? Even now, the British visitor to Australia is moved by the immensity of its war memorials, and by the thought of hundreds of thousands of young men bearing arms loyally for

the land from which, in many cases, their ancestors had been transported in chains. Australia sent more than four hundred thousand
men to fight alongside the British in the First World War, and nearly
a million in the Second. What explains such fidelity?

The answer has a great deal to do with the lessons learned in
London after the loss of the American colonies. The men and measures associated with Lord North's ministry had been utterly discredited. Almost no one was now prepared to argue that British subjects
overseas should have different political rights from those who remained in the British Isles.

It was understood that there were difficulties posed by distance,
hostile terrain, and relations with indigenous peoples. Nonetheless, a consensus soon formed in Westminster that British colonies
should be made ready for what was then known as "responsible
government"—meaning full parliamentary autonomy under the
Crown.

The early Australian population took eagerly to self-government.
The fact that it was largely made up of lawbreakers was no bar to its
political development. On the contrary, criminals of that era were
impressively enterprising. They were not part of an underclass of the
sort we know today. Indeed, the severity of their sentences strikes us,
at this distance, as shocking. English-speaking society, as we have
already noted, was exceptional in its reverence for property, and this
reverence was reflected in the criminal justice system. Thousands
were condemned to seven years of penal servitude for what we
would nowadays call shoplifting. Others were transported as political radicals, naval mutineers, or Irish nationalists.

The first eleven transport ships left Portsmouth, England, for
Botany Bay in 1787, carrying 696 prisoners and 348 officials, sailors, and marines, as well as stores, medical supplies, handcuffs, and
chains, and a prefabricated house for the governor, complete with
glass windowpanes. The last convicts arrived in Western Australia

in 1868. Between those two dates, 164,000 men, women, and children (many conceived en route) were shipped from the British Isles to Australia, along with a handful of troublemakers from Canada, India, and the Caribbean.

The journey was hellish, and many died during the passage. Conditions upon arrival were abominable, with vicious discipline on top of eerily unfamiliar flora, poisonous fauna, a harsh climate, and chronically hostile aboriginal tribes.

Yet the colony prospered. Prisoners were granted thirty acres of land on the completion of their sentences, and it was not long before a local economy began to flourish, based initially on sheep farming but soon sustaining all manner of shops, inns, and other secondary industries. The convicts, most of whom had come from the squalor of the slums or—even worse—the utter wretchedness of rural poverty, found themselves in a place where it was relatively easy to become a landowner. They grew into their new homeland, becoming rugged individualists.

Their success was bought at the expense of the indigenous inhabitants. Nowhere was the technological gap between settler and aboriginal wider than in Australia, whose autochthonous culture had not progressed beyond the Paleolithic stage. The convict ships brought catastrophe to the indigenous peoples: loss of hunting grounds and water, virtual enslavement on sheep and cattle stations, and, worst of all, unfamiliar pathogens. Although politicians in London periodically decreed various schemes aimed at assimilating aboriginal Australians into British society, local whites dismissed such schemes as ludicrously sentimental. In any event, no Westminster legislation could protect the natives from measles, tuberculosis, or smallpox.

As the settlements spread along the Australian coastline, a society began to develop that, as in North America, exaggerated the traits contemporary Europeans associated with the British. The

British had, historically, been remarkably ready to defy their rulers. They had elevated the individual not only over the state but also over the family unit. They had valued independence and self-reliance.

Australians took these characteristics much further. Like the American colonists, they had no territorial aristocracy on their soil. Land was plentiful, and the frontier was expanding. British emigrants were, so to speak, allowed to be themselves. The great red landmass became, not a spillover Britain, but an intensified Britain.

Any visitor to Australia is struck by the endurance of these characteristics: informality, bloody-mindedness, individualism, self-reliance. The Australian writer Sally White produced a short guidebook, explaining the national character for the benefit of foreign students: "Australians don't respect people just because of their role in society or their birth. As long as someone's behavior doesn't interfere with another person's activities or beliefs, Australians are tolerant and easy-going."

Here, in short, is Mill's libertarian philosophy made flesh. Which is precisely what we should expect to find. For, although Australians who can boast convict ancestors tend to be enormously proud of them, the felons were outnumbered by the adventurers who came looking for gold. As with the convicts, the prospectors were disproportionately male, which served to exaggerate still further the individualism that visitors and Australians themselves associate with that country's culture.

The first major gold strike was in 1851. Over the next twenty years, Australia's population increased from 430,000 to 1.7 million. And the discoveries kept coming. Throughout the second half of the nineteenth century, mines were opened across New South Wales, Victoria, and Queensland. The last major strike was made in Kalgoorlie in 1893, setting off Western Australia's gold rush. The prospector was a distant kinsman of mine, an immigrant from County Clare in Ireland named Paddy Hannan. Like the thousands who had crossed half the world to seek their fortunes, he was an entrepreneur: an adventurer prepared to take risks in pursuit of his fortune. Australia was not peopled by men who looked to the government for solutions.

Not unnaturally, Australians evolved a cartoon version of the place they had left behind. Britain, in their fancy, seemed class-ridden, pernickety, obsessed with the niceties of etiquette. And, from an Australian perspective, perhaps it was, though few non-Anglosphere visitors saw it that way.

Yet hardly any Australians questioned their links with Britain, which, until well into the twentieth century, was widely referred to as "home," even by those who had never set foot there. They knew that, while there were differences and rivalries, these existed within the continuum of a common political culture. A traveler from one country to the other, though he had crossed half the world, would find that the courts functioned the same way, the parliaments looked and behaved alike, and the unwritten rules that make any society function were compatible. It was these affinities, rather than those of sport, television, food, and so on, that were the core of national identity, and later generations of migrants to Australia, first from southern Europe and then from Asia, eagerly adopted them.

Until the 1960s, a majority of Australians identified themselves as British as well as Australian. This identification later waned, partly through the passage of time, partly because of large-scale migration from elsewhere, and partly because of the United Kingdom's calamitous decision to join the European Economic Community (now the European Union) in 1973, replacing the preferential trade relationship that had existed between the two countries with Europe's Common External Tariff.

Even now, many Australian visitors are dismayed, on arriving at British airports, to find that they must line up along with the rest of the world while the citizens of countries against whom they fought alongside Britain come in through the EU channel. "There were no bloody queues at Gallipoli," they complain, recalling the disastrous campaign to break through to the Black Sea in which so many Australian and New Zealand soldiers died in 1915.

At the time of the Gallipoli fiasco itself, it did not occur to Australian and New Zealand soldiers to draw a distinction between themselves and their British comrades. Indeed, the British suffered a very slightly higher casualty rate, proportionately, than the Australians did. This point is worth making, if only because of the mythology that has grown up around that campaign, especially since the 1981 film *Gallipoli*, starring Mel Gibson, which suggested that the Australian forces at the Straits suffered unnecessarily under the command of callous, tea-drinking British officers.

Gallipoli was indeed a critical moment in the development of Australian (and New Zealand) nationhood, and the commemorations of ANZAC Day surpass in scale the equivalent Remembrance Day/Armistice Day/Veterans Day ceremonies held throughout the Anglosphere on November 11. But the idea that its significance lay in a loosening of ties to Britain would have struck almost all the combatants as preposterous. To be sure, the Gallipoli campaign was badly planned and ineptly executed. But few at the time doubted that the English-speaking peoples were fighting side by side against tyranny.

Indeed, the move to independence for the colonies with large British populations did not primarily come from the colonists, as it had in 1776. Precisely because the lessons of the Second Anglosphere Civil War had been learned, the London authorities sought to remain always one step ahead of the colonies, offering more autonomy than had yet been requested.

The Statute of Westminster, which removed the British Parliament's last remaining powers of legislation over the Dominions, was passed in 1931. But it was not ratified by South Africa until 1934, nor by Australia until 1942, nor by an especially reluctant New Zealand until 1947. In Canada, arguments between the federal and provincial authorities delayed complete adoption until 1982.

The evolution of these great nations from dependencies into sovereign allies had been the vision of most nineteenth-century British policy makers and, in the case of Australia, it was a resounding success. In 2000, to mark the centenary of the law that had provided for the federation of the six self-governing colonies as the Commonwealth of Australia, five Australian prime ministers—Sir John Gorton, Gough Whitlam, Malcolm Fraser, Bob Hawke, and John Howard—came to the House of Commons, the place where the bill had been passed. They returned a few months later to mark the centenary of the federation itself with a state banquet at the Guildhall and a service at Westminster Abbey attended by the queen, whom their countrymen had voted to keep as head of state in a referendum the previous year. There were, to be sure, ceremonies in Australia, too. But it is significant that such an important part of the commemoration of Australian nationhood should have taken place in the United Kingdom.

Returning to London in 2003, John Howard attended the opening of a memorial to the hundreds of thousands of Australians who had fought for the values of the English-speaking peoples in the two world wars. I happened to be walking past as the ceremony took place. More than two hours later, I returned the same way, to see the amiable, unflashy, supremely successful Australian leader still there, the cameras long departed, taking time to talk to the many London-based Australians who wanted to shake his hand.

Howard, the first serving head of government publicly to use the word *Anglosphere*, was in no doubt as to what that memorial marked. The English-speaking peoples had twice stood almost alone in defense of liberty. In his eyes, they still did. As he was to tell an American audience in 2010:

I have found in my political life that the instinctive familiarity
and closeness of our societies is a quite remarkable thing and it's
perhaps nowhere better demonstrated at the highest levels of government than in the fact that without question, the single closest
intelligence-sharing arrangement that exists anywhere in the world
is the intelligence-sharing arrangement between the five members
of the Anglosphere—between the United States, Australia, Great
Britain, Canada, and New Zealand. When you think of the importance of timely intelligence in the fight against terrorism, it's a
remarkable tribute to the faith that we have in the integrity and
the reliability and the importance of that relationship.

In the same remarks, Howard spoke warmly of India having Anglosphere characteristics. How did he define the Anglosphere? By its
attachment to personal liberty, the common law, representative government, and—critically, in his view—its readiness to deploy pro-
portionate force in defense of those values. "It's a very long and rich
heritage in the defense of freedom," he concluded. And he was right.

LOYAL SHE REMAINS

Paradoxically, it was Canada—settled by those Americans who left
everything behind rather than forsake their allegiance to the British Crown—that gave nineteenth-century colonial policy makers
the greater headache. Australia's rugged individualists were happy
enough to accept British nationality. Yet, in 1837 and 1838, revolts
broke out among the English-speakers of Upper Canada (some of
whom sympathized with American republicanism) and the French-
speakers of Lower Canada (a few of whom sympathized with French
republicanism, and many more of whom simply disliked living under
British rule).

There was some excitement in the United States, which had invaded Canada at the beginning of the revolution, and again in 1812. Perhaps, at long last, George Washington's dream of a pan-continental republic might be realized.

This time, though, the London authorities did not react with Lord North's spectacular combination of petulance and high-handedness. Even as they moved to put down the risings, the authorities indicated that they were prepared to address the colonists' legitimate concerns.

The Whig government in Britain offered Canada a deal not unlike that which Pitt the Elder had wanted to offer America sixty years earlier: parliamentary self-government under the Crown. Some of the ringleaders of the Canadian revolt were transported to Australia, and a few were hanged. Yet, in passing sentence, Chief Justice Sir John Robinson offered an essentially Lockean justification for the punishments. The rebels, he said, had every right to espouse republicanism; but, in taking up arms against the legitimate authorities, they had endangered the liberty and property of their neighbors, and thus had broken the social compact.

The Earl of Durham, a languid but surprisingly militant Whig known as Radical Jack after his campaign for a wider franchise, was sent out to attend to Canada's grievances. He recommended the unification of Upper and Lower Canada (which happened) and the political and linguistic assimilation of the Francophones (which didn't). He also proposed responsible government, which subsequently came to colony after colony, starting in Nova Scotia (1848) and Prince Edward Island (1851), passing through New Zealand (1856) and South Australia (1857), and eventually reaching Western Australia (1890) and Natal (1893).

British lawmakers no longer wanted permanent control over their overseas kindred; rather, they wanted to retain their good-will. In 1891, a deputation from the Imperial Federation League approached the prime minister, Lord Salisbury, with the idea that the British colonies should copy Germany's Zollverein (customs union). The Tory patriarch, who had once supplemented his income by reviewing books in German, responded from the depths of his great beard that a Zollverein was all well and good, but what he was really after was a Kriegsverein (military union). He got it, too, with the happiest of consequences for the human race.

Canada initially looked like an outlier in an Anglosphere increasingly defined by Whig and liberal principles. It had, after all, been founded in a rejection of those principles. After 1783, almost no one in the United States dreamed of calling himself a Tory; but the word remained, and remains, current north of the forty-ninth parallel. At first, it applied to Loyalist exiles and their descendants. Then it came to mean those who looked to Britain rather than the United States in foreign policy. Now it is shorthand for supporters of the Conservative Party.

Canada's Toryism was bolstered by the Catholic and seigneurial culture of Quebec. Although the French-speakers were no great lovers of Britain, they had even less time for the ideals of the American revolutionaries. Some American exiles did indeed hope that their new homeland would develop along Tory lines, with a strong episcopacy and perhaps a colonial aristocracy. Indeed, all Loyalists had been granted a hereditary title by Guy Carleton, now Lord Dorchester and the governor-general of British North America, in 1789:

> Those Loyalists who have adhered to the Unity of the Empire, and joined the Royal Standard before the Treaty of Separation in the year 1783, and all their Children and their Descendants by either sex, are to be distinguished by the following Capitals, affixed to their names: U.E., alluding to their great principle, the Unity of the Empire.

Though many Canadians are entitled to those letters today, few bother. For Canada, despite its Tory beginnings, soon became as meritocratic and individualist a society as any on the planet. This development owed something to the testing nature of a frontier society, and something to the fact that the Quebecois and the United Empire Loyalists were swamped numerically by waves of English Protestants, Scottish and Ulster-Scots Presbyterians, and German and Scandinavian Lutherans. Mainly, though, it was simply part of a general trend throughout the English-speaking world. Conscious of their history, most Canadians remained loyal to the Crown, but would have been as outraged as the most radical supporters of Wilkes or Paine had the monarchy become functional rather than decorative.

Nor should the Toryism of the defeated Loyalists be mistaken for statism or servility. Many had left the infant American Republic not because they were unthinking Royalists, but because they feared that mob rule would lead to socialism. As Daniel Bliss, a Massachusetts exile who later became chief justice of New Brunswick, put it, "Better to live under one tyrant a thousand miles away than a thousand tyrants one mile away."

Visiting Canada while preparing this book, I found several of my prejudices challenged. I had had the country all wrong—and so, it struck me, had one or two Canadians. I grew up thinking of that vast land as a sort of touchy-feely version of the United States, obsessed with multiculturalism and the supremacy of the United Nations. In foreign policy, as in domestic, Ottawa seemed to owe more to Scandinavia than to the Anglosphere. The Canadian writer Robertson Davies—for my money the finest novelist of the late twentieth century—has a Swedish character in one of his trilogies declare herself quite at home in another socialist monarchy.

In an episode of *The Simpsons*, Homer tells the queen: "I know we don't call as often as we should, and we aren't as well behaved as our goody-two-shoes brother Canada. Who by the way has never had a girlfriend. I'm just saying." Such is the power of American popular culture that even some Canadians started to see themselves in such terms.

Only after it had passed did I see what an aberration the "goody-two-shoes" phase had been. Before the ascendancy of Pierre Trudeau in 1968, Canadians prided themselves on being hardier and more independent than their American neighbors. Their immigration policy was based on maintaining more attractive tax rates than the United States, so as to compensate for the rougher climate. They avoided FDR's spending splurge just as they avoided Obama's. In consequence, their combined provincial and national administrations account for a lower percentage of national gross domestic product than do the combined state and federal governments south of the border.

Of course, deep down, we Britons knew that there was something bogus about the goody-two-shoes, officious Canada that Mark Steyn calls "Trudeaupia." We remember Canadians as allies on the battlefield. Any British veteran will tell you that, in the two wars, Canadian soldiers had a reputation for grim and terrible courage. Dwight D. Eisenhower used to remark (in private, obviously) that, man for man, they were the finest troops under his command. At Vimy Ridge, Passchendaele, and the Somme, Canada lost nearly seventy thousand men out of a total population of seven million. In the Second World War, a further forty-five thousand Canadians perished in Italy and France. All for the sake of a mother country that, in most cases, they had never set eyes on. We Britons, of all people, should be delighted to find them back to their old ways.

For, truly, they are back. After more than thirty years of being a sort of outpost of Continental Europe in North America, Canada has convincingly rejoined the Anglosphere. Stephen Harper, who has been in office with an increasing share of the popular vote since 2006, is perhaps the most pro-British and pro-American leader in the world.

In one of his first speeches as prime minister, Harper spoke movingly to a London audience about how lucky his country was to have been founded in the British political tradition, and how the inherited folkright of common law was the basis of Canada's freedom. As a statement of historical fact, his proposition was hardly controversial. But it was a radical break with the line taken by previous Canadian leaders, namely that their country was a happy fusion of indigenous peoples, Acadians, illegal immigrants, and so on. Still, it went down well with Canadians—not least those from ethnic minorities, who backed his Conservative Party in unprecedented numbers.

Indeed, in the 2010 Canadian election, immigrants were more likely to vote Conservative than native Canadians. Think, for a moment, about how exceptional such an outcome is. In most of the world, newcomers vote overwhelmingly for left-of-center parties. There are plenty of reasons why. They are usually penniless when they arrive, and so gravitate to politicians who purport to represent the poor. They tend to live in districts represented by leftist politicians who, at local level, are the first to help them navigate the political system of their new nation. And, of course, left-wing parties see themselves as champions of all minorities.

The success of the Canadian Conservatives owed a great deal to their determined campaigning among ethnic minorities and new settlers. But it was assisted by the nonracial way in which English-speaking societies define themselves.

ANGLOSPHERE, NOT ANGLOS

The United Kingdom, itself a fusion of nations, was early in defining its nationality with reference to ideals, not race. This is not to say that racism was unknown in British society; far from it. But the law was color-blind, and nationality was not tied to birth or ancestry, as it was in most European states. In the years after the Second World War, the United Kingdom assimilated a large number of nonwhite immigrants, far earlier and more successfully than neighboring countries. The influx was not the result of a conscious decision to invite overseas workers; it happened automatically, because Britain afforded the same rights of nationality and residence to all its subjects, whether in Aden or Jamaica.

The core Anglosphere states are all now multiracial, yet they have been relatively undisturbed by major ethnic tensions. Only South Africa had a history of serious racial problems—a history that would unquestionably have been very different had British South Africans, rather than the Afrikaners, been the larger white population.

We now remember the end of apartheid as a liberation for black South Africans, which of course it was; we forget that it was seen as no less of a liberation by English-speaking whites. We forget, too, that the most serious rioting during the apartheid years was over the attempt in 1976 to impose the tongue of the Boers alongside English as a medium of instruction in schools. Black South Africans saw Afrikaans as a language of oppression, English as a language of opportunity.

Touring the Anglosphere in 2011 and 2012, I was struck again and again by the multiracial composition of the audiences I spoke to—especially in Canada, Australia, and New Zealand. These were self-selecting audiences: they had come to listen to me extol the alliance of English-speaking democracies. Yet they were ethnically representative of the local population. Though their grandparents might have come from Korea or Vietnam or Ukraine, they plainly felt that becoming Australian or Canadian didn't simply mean accepting jury trials and multiparty democracy and the sanctity of contract; it also implied a special relationship with other countries that elevated these values. There is poignancy in the fact that the core Anglosphere territory that has been least successful in inculcating that sense in immigrants is the United Kingdom itself.

Race is always a difficult subject for historians. A certain type of critic insists on judging past ages by contemporary standards, and thereby condemning Jefferson or Churchill or virtually any other historical figure by an act of, so to speak, retrospective jurisdiction.

There is no denying that the coming of English-speaking peoples was a disaster for many of the earlier inhabitants of the lands in which they settled. There was massive depopulation among the native peoples of North America and Australia, and the indigenous tribes of the Caribbean were almost extirpated. This was mainly a consequence of the viruses carried by the planters rather than of deliberate policy.

Nonetheless, the colonists' callousness strikes our generation as horrific. The Canadian-born cognitive scientist and popular writer Steven Pinker has argued persuasively that, over centuries, humanity has become less violent: deaths from homicide and wars keep falling as we widen the circles within which we recognize mutual obligations. Those circles, in the seventeenth century, were still relatively tight.

Listen to the way even the Whig heroes, men whom we have already encountered favorably in these pages, speak of those who were beyond the circles of reciprocity.

John Locke, the author of the still-dominant Anglosphere theory of government, argued that barbarians, like criminals, were not covered by the social compact, "and therefore may be destroyed as a *Lyon* or a *Tyger*, one of those wild Savage beasts, with whom Men can have no security."

"God hath consumed the natives with a great plague in those parts, so as there be few inhabitants left," noted John Winthrop, leader of the Pilgrim settlers in Massachusetts Bay, justifying the expansion of their colony. As for those who remained, "they enclose noe land, neither have they any settled habitation, nor any tame cattle to improve the land by, & soe have noe other but a natural right to those countries. Soe if wee leave them sufficient for their own use, wee may lawfully take the rest."

Sir Edward Coke, the great parliamentary leader whose works formed the basis of Anglo-American jurisdiction for two centuries, drew a distinction between Christian foreigners, with whom the English might treat under a moral and legal code understood by both parties, and barbarians: "All Infidels are in law *perpetui inimici*, perpetual enemies (for the law presumes not that they will be converted, that being *remota potentia*, a remote possibilitie) for between them, as with the devils whose subjects they be, and the Christian, there is perpetual hostility, and can be no peace."

It is hard to know the number of Native Americans who perished as a result of epidemics, loss of hunting grounds, and war. The anthropologist Henry Dobyns puts the total pre-Columbian population of the United States at 18 million, though other estimates range from 2.1 million to 7 million. What we do know is that the number had fallen to 750,000 by 1700, 600,000 by 1800, and 300,000 by 1900.

In Australia, an indigenous population of perhaps 750,000 at the time of the Botany Bay landing had fallen to 93,000 by 1900.

In New Zealand, too, the main killer of the indigenous peoples was disease. The second was the disequilibrium brought by British technology. Some Maori tribes acquired muskets, which turned the tribal wars—important in a culture that emphasized feats of courage and arms—into genocides for the tribes without equivalent weaponry.

Nevertheless, Britons in New Zealand were keen to assimilate the Maoris fully into local social and political structures. The fighting skills of the indigenous peoples were particularly admired. A Maori force—the Native Contingent—joined the Gallipoli offensive in 1915 and, in the Second World War, the Maori Battalion saw action in Greece, North Africa, and Italy. Maoris were quickly granted the same political and civil rights as settlers, including the vote.

The political culture and common-law heritage of New Zealand's settlers made it unthinkable for Maoris to be placed in a separate legal category: the fate of many native peoples living in European empires at the time. Almost all policy makers from the mid-nineteenth century onward aimed at full assimilation, and some went further. In 1903, the patriotic MP Sir William Herries looked forward to a time when "we shall have no Maoris at all but a white race with a dash of the finest colored race in the world."

Consider that sentiment, not by the standards of our present political dialogue, but in the context of its time. The idea that a native population would integrate on terms of complete equality with white settlers would have been unthinkable in almost any other contemporary colonial empire.

It was not that English-speakers were more racially enlightened than Belgians or Italians or Germans. Rather, their attitudes were shaped by their institutions. Yet again, the true hero of our story turns out to be the common law.

In 1772, the English legal system had distinguished itself with a ruling that elevated and ennobled every British subject. A customs officer named, unhappily, Charles Stewart, had been stationed in Boston, where he had purchased a black slave named James Somersett. On their return to England, Somersett escaped and was recaptured. Abolitionist campaigners argued that the ownership of one human being by another was automatically dissolved on British earth, since no parliamentary statute provided for slavery. The case was unprecedented, and the judge, Lord Mansfield, therefore sought to "discover" the common-law principles that applied to it. What he discovered brooked no argument:

> The state of slavery is of such a nature that it is incapable of being introduced on any reasons, moral or political, but only by positive law [a parliamentary statute], which preserves its force long after the reasons, occasions, and time itself from whence it was created, is erased from memory. It is so odious, that nothing can be suffered to support it but positive law. Whatever inconveniences, therefore, may follow from the decision, I cannot say this case is allowed or approved by the law of England; and therefore the black must be discharged.

Six decades were to pass before slavery was finally eliminated throughout the British Empire, and another three before it was extirpated in the United States. Yet all Anglosphere peoples can take pride in the vigor with which the abolitionist campaign was waged.

Slavery violated the principles that the English-speaking peoples regarded as peculiarly theirs. It was, obviously, incompatible with personal liberty, and with the free exchange of labor on which open markets rested. It was especially abhorrent to Whig-Protestant sensitivities, and the abolitionist movement was led by evangelical and Nonconformist church groups. Much of the campaigning was remarkably modern in its tactics. The famous Dissenter Josiah Wedgwood, who created the pottery works that bear his name, produced medals with the image of a slave in chains asking, "Am I not a man and a brother?" The medals became fashion items: ladies would wear them as pendants or in hairpins, and larger versions would hang on walls.

In 1807, in response to popular demand, Parliament outlawed the traffic in human beings and began a long and grueling war of attrition against the slave runners. It was an early example of a moral foreign policy, powered by pressure from the electorate. The United Kingdom persuaded or bullied other European states, as well as African kings, into agreeing to halt the transatlantic trade, and the Royal Navy was deployed against the slavers. Between 1808 and 1860, some 1,600 ships were seized, and 150,000 Africans freed.

Britain derived neither profit nor diplomatic advantage from these seizures. Indeed, it diverted resources to the antislavery campaign even while waging a deadly war against Napoleon.

These facts are worth reprising, because slavery is still thrown in the face of the Anglosphere peoples by their detractors. It cannot be stressed too often: the institution existed in every age, in every society, on every continent. What distinguished the English-speaking nations was not that they practiced slavery, but that they crushed it.

Some critics, looking for a peg on which to hang their prejudice, object to the fact that slave owners were compensated at the time of emancipation in 1833. A British leftist paper, the *Independent on Sunday*, referred in 2013 to these payments as "Britain's colonial shame." The head of the United Kingdom's Equality and Human Rights Commission, an agency established by the Blair administration, called it "the most profound injustice that probably you can identify anywhere in this country's history."

Really? The fact that people were prepared to *pay* to abolish the monstrosity of slavery is surely a cause for satisfaction rather than shame. It is one thing to say, in the abstract, "slavery is a bad idea," and quite another to say, "slavery is so wicked that I am prepared to make a personal sacrifice to help do away with it."

When, thirty years later, the United States got around to emancipation, no compensation was paid. Instead, a terrible war was fought, whose legacy of racial bitterness endured for another century and more. Yet when Congressman Ron Paul suggested that it might have been better for everyone had the Americans adopted the British approach, buying out the slave owners peacefully, he was pilloried.

It is true that, with a handful of exceptions, the slaves themselves received no compensation. This was a terrible wrong. But that wrong does not invalidate the policy of purchased manumission.

Of course, if your starting point is that Britain and the United States were evil and oppressive colonial powers, you will find something or other to complain about. The absurdity of the whole debate, though, is that we are all descended from slaves; from slave owners, too, come to that. It could hardly be otherwise, human history being what it is.

Slavery was common to all agrarian societies. It persisted in every early civilization: in Ur and Sumer, in Egypt and Persia, in the Indus Valley and in Xia Dynasty China. It survived through the classical age, and into the medieval period.

Slavery was endemic in African and Arab societies. Between 11 and 17 million people were taken from Africa by Muslim slavers between the seventh and nineteenth centuries. In the New World, too, slavery existed from the earliest moment of human settlement. The Mayans, Aztecs, and Incas all practiced it.

Although slavery sometimes had an ethnic basis, it was no great respecter of race. Muslim slavers traded in Christians: Georgians, Circassians, Armenians, and others. Christians, for their part, enslaved Moors: as late as the sixteenth century, hundreds of thousands of Muslim slaves toiled on Spanish plantations. On the eve of the American Civil War, there were three thousand black slave owners in the United States.

We are, in other words, all in this together. Everyone on the planet is descended from the exploiters and the exploited. And that, surely, is what makes the arguments over guilt and apologies and

reparations so silly. We can all agree that slavery was an abominable crime. From a contemporary perspective, it seems unbelievable
that otherwise humane societies tolerated it. It is understandable
that we still flinch with revulsion at the thought that our ancestors
engaged in the practice. But then, so did the ancestors of every
other human being on earth. If we are to single out Anglosphere
nations in this context, it can only be to note their unique dedication to eliminating the evil.

None of this is to minimize the participation of English-speaking
peoples in the slave trade, nor its longevity in the American South.
It is simply a plea for perspective. Judged against contemporary societies, rather than by present standards, the Anglosphere was exceptional in its attachment to liberty. If we are to reel off the familiar
charges made by anti-Americans—the valuation of slaves at three-
fifths of a freeman, the continuation of segregation after emancipation, and so on—it is only fair to give the other side of the story, to
recall the energy of the abolitionist campaigners and the price they
were prepared to pay, including death on the battlefield, to vindicate
their beliefs.

To accuse the United States of hypocrisy, as so many of its domestic and foreign critics do, is entirely to miss the point. Hypocrisy means failing to live up to our ideals—something that, being
human, we all do. A society without hypocrisy would, by definition,
be a society without ideals.

To point to the gap between theory and practice in the United
States is nothing more than to note that the country is inhabited
by human beings. Ideals are supposed to serve as an incentive to
try harder and, in the United States, they did. It was precisely because Americans were aware of the gap between theory and practice that they poured so much energy, first into abolishing slavery,
and then into abolishing legal and semilegal racial segregation. It
is striking, in the speeches of Martin Luther King Jr., how rarely

he appealed to universal ideals and how often he called on the United States to live up to its ideals. "We will reach the goal of freedom in Birmingham and all over the nation," he wrote in 1963, "because the goal of America is freedom." The system, you might say, worked.

Much the same is true of the other Anglosphere lands. In his history of the British Empire, Niall Ferguson makes a point that, while obvious, is almost never allowed. It is certainly true that, along with the benefits of colonial rule—the roads, the clinics, the courts—there were costs. But, for almost every colonized people, the alternative to British jurisdiction was not unmolested progress toward modernity, but conquest by someone else—the French, the Germans, the Turks, the Italians, the Russians, the Japanese, or, worst of all, the Belgians.

There is little doubt that, given the alternatives, Anglosphere control was preferable. British rule—or American rule in the Pacific—was at least aimed at eventual democratic autonomy. In most colonies, independence was in time achieved without a shot being fired in anger.

When, for example, the Malayan Federation became independent in 1957, the chief minister, Tunku Abdul Rahman, presented the deeds of the former governor's residence, Carcosa Seri Negara, along with forty acres of prime land around it, to Britain "as a token of the goodwill of the Malayan people to Her Majesty's Government." An anticolonialist minister protested that, if the British were presented with so fine a palace, future generations wouldn't believe that the Malayans had had to struggle for independence. There was an awkward silence around the cabinet table: there had been no struggle.

In the settler colonies, the places with large British-descended populations that form the core Anglosphere today, indigenous peoples were eventually given an effective choice between assimilation or life on protected reservations. In the later colonies, especially in Africa, where British rule was from the first seen as provisional, sincere attempts were made to create the legal and physical infrastructure that would prepare local people for eventual statehood—although, in the event, Britain was so financially exhausted by the Second World War that it ended up scrambling away hastily and prematurely from its African possessions, with unhappy consequences.

Between these two extremes lies the territory around which the empire was built, a territory with a population that was not British by origin, but which has adopted the democratic, common-law, and individualist values of the Anglosphere, a country on whose orientation the present century turns.

MACAULAY'S CHILDREN

Is India primarily an Anglosphere democracy or an Asian superpower? It may be the single most important geopolitical question of our age.

There are many educated Indians who speak English at home, are conversant with English literature and philosophy, and find the question mildly insulting. Here, to pluck an example more or less at random, is a miffed response by the Takshashila Institution to a debate among Australian politicians that had left their country off a list of Anglosphere states:

> India is functionally Anglophone, with English not just the language of the central government and universities, but also formally one of its national languages. Informally, it is the medium both of nationwide intellectual exchange and, to a lesser degree, business. India also shares some of the other socio-cultural characteristics

that unite much of the Anglosphere, including a legal system grounded in common law, representative democracy, and religious and ethnic pluralism. Few major countries outside the United States, Britain and the former British dominions check all those boxes.

True. Nor are there many countries in that part of the world with armies that are wholly under civilian control. Nor, for that matter, where power passes peacefully from party to party at elections without the losers being exiled or shot. It is also worth remembering that India has the third-largest Muslim population on earth, yet has had remarkably few problems with jihadi extremism.

There is, though, far more to the Anglosphere than language, common law, and religious pluralism. The Anglosphere is a compound of several metals, some of them mined in the Middle Ages and forged in the furnace of the seventeenth- and eighteenth-century civil wars. The resulting alloy was carried by English-speakers to North America, parts of the Caribbean, the Cape of Good Hope, Australia, and New Zealand. But how does India, with its very different early history, fit in? Can we press that vast land with its many overwritten narratives, that palimpsest, as Jawaharlal Nehru called it, into the Whig-Tory arguments that defined the political orientation of the other Anglosphere states? Can we identify a distinctively Anglosphere approach to individualism and enterprise?

One way of answering the question is to imagine that France's Compagnie des Indes, rather than the East India Company, had prevailed in the struggle for supremacy there. The two companies had similar names and a similar purpose: trade. But they were very different in structure. The Compagnie des Indes had been established by the state. It was funded, ultimately, by the French monarchy, and its directors were government-appointed. The East India Company, by contrast, was a private venture, answerable to its shareholders.

Not surprisingly, its agents had a strong commitment to property rights and free contract—a commitment that to this day sets India apart from many nearby states.

During much of the period when the English-speaking peoples were asserting their exceptionalism, India was subject to the same Whig-Tory tussles as the rest of the Anglosphere. It is perfectly true that these tussles were taking place largely within a British-born elite. Throughout the period of British rule, most Indians did not speak English, and John Locke's treatises on government were as remote to them as distant stars. Then again, the same is true of most of the core Anglosphere nations. Politics is always a minority pursuit. Very few of the Americans who took up arms in the Patriot cause had heard of Locke either; but his teachings affected their lives, and those of their children, regardless. Ideas have consequences.

We have encountered Lord Macaulay as the greatest of all the Whig historians, whose chronicle of the Glorious Revolution has not been out of print through more than one and a half centuries. He was also a politician and Indian administrator, who lived between 1834 and 1838 in Calcutta, then the capital of British India. He foresaw that, as British habits—by which he meant science, modern medicine, representative government, and personal liberty—penetrated the Indian population, the people would begin to aspire to independence. He welcomed the prospect for reasons that, in retrospect, are hard to dispute:

> To the great trading nation, to the great manufacturing nation, no progress which any portion of the human race can make in knowledge, in taste for the conveniences of life, or in the wealth by which those conveniences are produced, can be a matter of indifference. It is scarcely possible to calculate the benefits which we might derive from the diffusion of European civilization among the vast population of the East. It would be, on the most

selfish view of the case, far better for us that the people of India were well governed and independent of us, than ill governed and subject to us; that they were ruled by their own kings, but wearing our broadcloth, and working with our cutlery, than that they were performing their salaams to English collectors and English magistrates, but were too ignorant to value, or too poor to buy, English manufactures. To trade with civilized men is infinitely more profitable than to govern savages.

Macaulay's vision animated many British administrators. They wanted to prepare India for eventual autonomy. Schools that offered a British-style education became popular, especially in Bengal. The civil service was opened to Indians, by competitive examination. Railways—"the greatest destroyer of caste," as one enthusiast called them—girded the country.

As education expanded, increasing numbers of Indians began to participate in government. Legislation was passed establishing elected Indian local councils in 1908. In 1934, the India Bill provided for autonomy on the Canadian or Australian model. Congress—the party that led India to independence and was in power for most of the succeeding period—had been founded by a British liberal.

Though Whigs generally favored eventual self-government for India, they were the more culturally chauvinistic party. They saw Indian religions as superstitious and retrograde, and hoped that they would wither in the face of Western rationalism. "No Hindu who has received an English education ever remains sincerely attached to his religion," asserted Macaulay in 1837, with that smugness that was his chief fault. But the British authorities were also prepared to intervene directly against practices which they found abhorrent. There is a famous story told of Sir Charles Napier, a tough career soldier, being approached by a delegation of Hindu priests to complain about the British proscription of sati, the custom by which widows were burned alive on their husbands' funeral pyres. According to his brother's account, Napier replied:

> Be it so. This burning of widows is your custom; prepare the funeral pile. But my nation has also a custom. When men burn women alive we hang them, and confiscate all their property. My carpenters shall therefore erect gibbets on which to hang all concerned when the widow is consumed. You may then follow your custom, and we shall follow ours.

As in the Americas and in Australia, there were tensions between the London authorities and local administrators. Parliament feared that British officials would go native—would, in other words, lose their belief in independence and liberty, and become as lethargic and despotic as Orientals were supposed to be. The men on the ground, the East India Company officials, scorned what they saw as the softheadedness of those who had no idea of India.

The awkward fact, as both sides were aware, was that India had been acquired by a series of acts of looting and theft. What had begun as a commercial enterprise turned into an administrative one, as local rulers invited East India Company troops to support them in disputes in exchange for payment, and as the British rushed to control princedoms before the French could beat them to it. Almost without intending it, the East India Company ended up running vast swaths of territory, usually in the name of local princes.

The man who had done most to effect these stunning conquests, Robert Clive, was contemptuous of the rajahs whom he had defeated: "Indoostan was always an absolute despotic government. The inhabitants, especially in Bengal, in inferior stations are servile, mean, submissive and humble. In superior stations they are luxurious, effeminate, tyrannical, treacherous, venal, cruel."

Clive was the first of many East India Company men to

have been hauled before Parliament to explain what looked like straightforward abuse of office: by a series of deals and military alliances, he had diverted the tax revenues of large parts of Bengal to his own pocket.

His characterization of the East as sybaritic and autocratic chimed with the prejudices of MPs. Indeed, it confirmed their fear, namely that, instead of elevating Indians, men like Clive were sinking to their level, losing their Protestant moral compass and forgetting that they were Englishmen. The Whig leader, Lord Rockingham, accused the company of "rapine and oppression" in Bengal, and one of his MPs—the same General John Burgoyne who was later to prove so reluctant to prosecute the war vigorously against the American Patriots—recommended that the government there be brought into accord with "the principles and spirit of the British constitution."

Many Tories thought such notions naive. One such was Richard Wellesley, who served as governor-general of India between 1798 and 1805, and was the elder brother of the future Duke of Wellington. Wellesley was, like a lot of Anglo-Irish peers of his time, something of a snob. "I wish India to be ruled from the palace, not the counting house," he wrote, "and with the ideas of a prince, not those of a retail trader in muslins and indigo."

His more famous brother, who had served in India as a soldier while Wellesley was governor, was just as dismissive of the Whig notion that Indians might be habituated to British concepts of liberty. In 1833, the hero of Waterloo opposed the abolition of slavery in India, arguing that Britain must "uphold the ancient laws, customs and religions of the country."

It was the religious issue that was to provoke the strongest challenge to British rule: the Mutiny of 1857. The evangelical activities of British missionaries had been causing increasing anxiety among Hindus and Muslims. A rumor circulated among the Indian soldiers in British service, the sepoys: it was said that the paper on the cartridges for their Enfield rifles was greased with both cow and pig fat—making it unclean to both major religions.

The uprising that followed stunned the British authorities, and ended with terrible reprisals against the mutineers, some of whom were fired from cannons. As usually happens at moments of shock, each side reverted to its prejudices. Tories blamed the insurrection on Whig meddling, especially the lack of respect shown to traditional religious leaders. Whigs, for their part, blamed the venality of the East India Company, which was promptly nationalized. India at last passed formally under the jurisdiction of the British government.

Nationalization, however, did not end the argument between the modernizers and the "Toryentalists." Lord Lytton, the Tory viceroy who organized the first Delhi durbar, believed to the end that British principles of meritocracy and liberty could not be transplanted to Indian soil: "I can imagine no more terrible future for India than that of being governed by Competition Baboos."

The babus constituted the professional class, mainly in Bengal, which had been called into existence by the education reforms. Rational, English-speaking, and true to their schooling, they tended to favor Indian autonomy—though, as their critics were quick to point out, their background had sundered them from the rural masses.

Unlike the babus, the maharajahs were supposed to have natural authority. Lord Curzon, who had organized the second durbar, told a London audience at the end of his term that he wanted the native princes to be seen "not as relics, but as rulers; not as puppets, but as living factors in the administration."

In the end, though, Britain was driven by the logic of its own political culture to accept Indian independence. Yet again, Hollywood has created myths—this time, the myth of an intransigent Britain hanging on until kicked out. In fact, Home Rule had been granted in principle thirteen years earlier, though its implementation was delayed by the Second World War. A small number of extremists threw in their lot with the Japanese invaders, who recruited among captured prisoners. But the broad mass of Indians gladly took up arms for the Anglosphere. Whereas the Japanese-backed Indian National Army numbered forty thousand, nearly 2.5 million Indians fought for the Anglosphere in Europe, Asia, and Africa: the largest volunteer army in history. Though most of these volunteers favored Indian autonomy, they did not hesitate to defend the values of the Anglosphere over those of the Axis. India, in their minds, was already becoming the free, law-based democracy that it became in fact a few years later.

Mohandas Gandhi himself, who enjoyed and still enjoys almost saintly status among his countrymen, was in no doubt that Britain would ultimately live up to its principles. In his early campaigns for the rights of Indians in the Transvaal, he had discovered a difference between the attitude of the British and that of the Boers. As he told a dinner of the Madras Bar Association in 1915:

> As a passive resister, I discovered that I could not have that free scope [in the Boer Republic] which I had under the British Empire. I discovered the British Empire had certain ideas with which I had fallen in love, and one of those ideas is that every subject of the British Empire has the freest scope for his energies and efforts.

He therefore had no hesitation in backing the United Kingdom in the war then raging in Europe. In 1918, as the final German offensive surged dangerously into France, he told Congress:

> India would be nowhere without Englishmen. If the British do not win, whom shall we go to claiming equal partnership? Shall we go to the victorious German or the Turk or the Afghan for it? We shall have no right to do so. The liberty-loving English will

surely yield when they have seen that we have laid down our lives for them.

It wasn't quite that simple. There were disagreements over precisely what "equal partnership" would mean. There were native princes who, fearing for their position in a democratic India, wanted the British to remain. And there was the religious divide, which in the end proved decisive and resulted in the partition of India and the creation of an independent Pakistan. All these things delayed the independence process, and Indians had laid down their lives for the Anglosphere a second time before independence was agreed. An extraordinary 2.4 million Indians fought for the English-speaking peoples against fascism.

Until the last day of British rule, several Tories, including Winston Churchill, remained convinced that India could never be a successful democracy. These men were, by and large, heirs to the tendency that had opposed the progressive extensions of the franchise in Britain itself, and their gloomy prognoses on both questions were proved wrong.

India has succeeded triumphantly as a democracy. The Bengali babus became more and more numerous—ceasing to be exclusively Bengali, ceasing to be babus, and becoming the world's largest middle class, English-speaking, consumerist, and democratic.

Macaulay is perhaps more famous for his poetry than for his history. But the single most significant act of his life was the decision to ensure that education in India would be in English. He made it for reasons that were culturally insensitive, arrogant, and chauvinist: the whole of Arabic and Sanskrit literature, he averred, was not worth a single shelf in a European library. Yet who can doubt that the decision has worked out happily for India, which now has the priceless advantage of speaking the global language?

Macaulay understood the limits of what he could achieve:

It is impossible for us, with our limited means, to attempt to edu-
cate the body of the people. We must at present do our best to form
a class who may be interpreters between us and the millions whom
we govern; a class of persons, Indian in blood and color, but English in taste, in opinions, in morals, and in intellect. To that class
we may leave it to refine the vernacular dialects of the country,
to enrich those dialects with terms of science borrowed from the
Western nomenclature, and to render them by degrees fit vehicles
for conveying knowledge to the great mass of the population.

Even today, there remains a vast gap between the "class of
interpreters" and the "great mass of the population." The Indian
bourgeoisie may be drawn into a common online Anglosphere conversation, but there are lakhs of villages where people would settle
for a steady electricity supply, never mind broadband. (Yes, lakhs: it's
a two-way street, the Anglosphere.)

Then again, how many people in the housing projects of Baltimore or the tenements of Glasgow have been drawn into an online
Anglosphere conversation? Politics, to repeat, is a niche pursuit. In
every country, an active minority ends up setting the tone.

The Indian middle class is growing with dizzying speed, and
the country is changing palpably. The writer Akash Kapur, who left
India for an American boarding school at the beginning of the 1990s,
felt the change when he returned. Indians were suddenly shopping at
Amazon, frequenting Starbucks, calling each other "dude."

Every newly independent state goes through a phase of exag-
gerating, or at least underlining, the things that make it different
from the former colonial power. In the days after the end of the
British Raj, this distancing took the form of embracing supposedly
traditional economic policies based on self-sufficiency—swaraj and
swadeshi—and of trying to promote Hindi as a national language.

But, as the last of the generation who campaigned for independence passes, and as the economy booms, India no longer has anything to prove. As an occasional visitor, I notice what Kapur noticed as a returning émigré. It's not simply that English is much more widely spoken; it's that travel between India and the diaspora communities has become routine. On my last visit to Madras, I asked directions from an exotic-looking woman in a gorgeous sari. "Ooh, I dunno, love," she replied, in broad Cockney. Like me, she was a tourist.

Communities of Indian descent remain in almost every corner of the Anglosphere, including those that British settlers evacuated long ago: Fiji, South Africa, Malaysia, East Africa, Hong Kong, the Caribbean, Australia, New Zealand, Singapore. There are half a million people of Indian heritage in Australia, a million in Canada, 1.3 million in South Africa, 1.4 million in the United Kingdom, 3.2 million in the United States—a visible and growing emanation of India's Anglosphere status. This vast, English-speaking community forms a halo around India proper.

As the colonial era passes from memory to history, identification with the Anglosphere is shedding any remaining association with colonial cringe or nostalgia. Apart from anything else, it is now more closely associated with American than with British popular culture.

Ultimately, though, all these considerations are secondary. The Anglosphere is based on a common affinity, a sense of unifying identity. Like almost all Britons, I feel such an affinity in India in a way that I don't in Europe (except, occasionally, in Scandinavia and the Netherlands—places that share so many Anglosphere characteristics including, now, the English language, that they might be considered as honorary members). It can reside in a shared book, a shared joke, sometimes just a shared look, but it's there.

There's an Indian restaurant in Brussels whose owners speak no English. Every time I go there, I feel that there is something wrong: the proprietors are delightful people, but I can't wholly repress the idea that they've swapped sides. I once asked a visiting Indian MP to join me there for dinner. "I hope you don't mind my saying this," I murmured after the waiter had taken our orders, "but I feel slightly unnatural having to talk to an Indian waiter in French." "*You* feel unnatural?" said my friend. "How the bloody hell d'you think *I* feel?"

Perhaps George W. Bush's single greatest foreign policy success was to draw India back into the alliance of English-speaking democracies when he accepted the nation's nuclear status in 2006. That relationship has been vigorously cultivated by David Cameron but neglected by Barack Obama. Fortunately, Indians seem prepared to wait for a different attitude from Washington. They are a patient and courteous people.

THE FAILURE OF OCEANA

Past attempts to draw the Anglosphere nations together were frustrated by four things. First, communications: no civilization has ever been so geographically dispersed. Second, politics: as long as people defined unity in terms of governmental control of territory, it was impossible to amalgamate territories whose political culture was based on local self-rule. Third, race: Victorian dreams of Anglosphere integration were caught up with then-current ideas of racial characteristics, and the implied exclusion of populations not of British ancestry—Quebeckers, Maoris, Africans and Afrikaners, Chinese, and, above all, Indians—made a common definition of nationality practically unachievable. Fourth, history: in two core Anglosphere territories—Ireland and India—there were independence campaigns that led first to a breakaway and then to the exaggerated rejection by the new state of many colonial-era attributes and trappings.

Only now, in the twenty-first century, have these objections ceased to apply. Distance was progressively conquered by steamboats, the telegraph, and jet travel, but not until the broad penetration of the Internet could people throughout the Anglosphere take part in a common English-language conversation.

Some leftist commentators have been unnerved by the development. In 2010, the *Guardian*, Britain's high-minded liberal newspaper, ran a column titled "Trapped in the Anglosphere," which bemoaned the fact that, since the Internet had become widespread, Britons had shown more interest in Australian and American politics than with what was happening in France or Germany:

> The online information age, which should, in theory, have been expected to facilitate greater mental and cultural pluralism and thus, among other things, greater familiarity with European languages and cultures, has, in practice, had the reverse effect. The power of the English language, at once our global gift and our great curse, discourages us from engaging with those outside the all-conquering online Anglosphere.

It was an unconsciously revealing moment. Not only did the old media expect Britain's elites to pursue European integration against the wishes of the majority. They also expected to dictate our international news priorities, and could not hide their annoyance that we, the ignorant rubes who make up the general population, were more interested in an Australian election where both party leaders had been born in Britain than in the local polls in Nord–Pas de Calais, which our Euro-elites had declared to be important.

The Internet has redefined people's relations with the state. The exclusivity that a nineteenth-century government used to enjoy in its relations with its citizens has been diluted. We don't simply move from country to country with unprecedented ease; we can increasingly do so virtually, without stirring from our armchairs. As we do so, we form online communities whose affinity is linguistic and political, and where we are often literally unable to detect racial differences.

The tendency to define the Anglosphere in partially ethnic terms, though always a minoritarian one, and short-lived at that, understandably created a backlash: those whose ancestors had come from elsewhere felt excluded. Only in the last generation have all the Anglosphere territories become multiracial to a degree that makes it impossible to see their political characteristics as the peculiar heritage of one ethnic group.

At the same time, the memory of anticolonial struggles is fading. India, having at first been keen to emphasize the things that made it different, is now at ease with its Anglosphere democratic and legal institutions. It no longer contrasts itself against the former imperial power by stressing its precolonial characteristics; rather, it contrasts itself against neighboring states by stressing its parliamentary and law-based political institutions. The English language is now unequivocally seen as the key to the modern world, not a legacy from the past.

Ireland, traditionally the other outlier, has long since passed its apastron, its moment of maximum orbital distance, from the rest of the English-speaking world. The dream that Éamon de Valera and so many of his contemporaries had of an Irish particularism based on autarky, the elevation of the Catholic hierarchy, and the restoration of the Irish language, now seems to belong to another age. At the same time almost no one in Great Britain still sees Protestantism as a component of national identity. As far as daily life goes—shopping, eating, sports, television—the peoples of the British Isles are arguably closer than ever before. A visitor from the other side of the planet would struggle to understand why feelings had once been so intense.

The Anglosphere is becoming a devolved, flexible, commonwealth of independent states, bound together not by governmental institutions, but by cultural, commercial, and family links. Only now, perhaps, do we understand that such a union requires no state ties.

This is an entirely new insight. Even during the brief period before 1776 when the Anglosphere was under one crown, there were various schemes for federation or confederation. Pamphleteers and politicians on both sides of the Atlantic proposed ways to bring the archipelago of British territories, haphazardly acquired, into some kind of constitutional order. A few proposed granting the colonies representation at Westminster. Many more envisaged a series of local parliaments, united only in matters of foreign policy and foreign trade. The Dissenting minister Richard Price, for example, wanted a series of independent colonies overseen by a senate that could act as "a common arbiter or umpire."

Adam Smith, without any expectation of persuading anyone, put forward a similar proposal for a "constitutional union of Great Britain with her colonies," with a weak confederal authority overseeing autonomous parliaments. The confederal authority should, he suggested, initially be based in London but, as the population of the Americas grew, it would doubtless migrate across the Atlantic.

All such schemes came up against the hard reality of nature. In an age when a transatlantic voyage lasted nine weeks, any imperial senate would be physically removed from, and thus almost immune to the wishes of, its electorate. The practical consequences of distance were revealed, in a slightly opéra bouffe way, during the War of 1812, whose only major engagement was fought after the peace had been signed, but before news of it had crossed the sea.

The impracticalities of a transoceanic imperium in the eighteenth century were adumbrated, with his customary eloquence, by Edmund Burke:

Three thousand miles of ocean lie between us and them. No con-
trivance can prevent the effect of this distance in weakening government. Seas roll, and months pass, between the order and the
execution; and the want of a speedy explanation of a single point is
enough to defeat a whole system. You have, indeed, winged ministers of vengeance, who carry your bolts in their pounces to the
remotest verge of the sea. But there a power steps in, that limits the
arrogance of raging passions and furious elements, and says, "So
far shalt thou go, and no farther."

In 1775, this was an almost unarguable objection. But there was
an implied corollary. If, at some future date, a way could be found
to overcome these distances, then the idea of some kind of political
federation might be revived.

Which, sure enough, is what happened. The invention of steam-
ships cut the time of an Atlantic crossing from sixty-six days to ten. But
the real breakthrough was the laying of undersea telegraph cables, the
first of which was unwound in 1858 between Valentia Island, County
Kerry, Ireland, and, on the other end, Heart's Content, Newfoundland. Two converted warships, HMS *Agamemnon* and the USS *Niag-
ara*, set out from opposite coasts, paying out the tarred, hemp-covered
copper wire as they went and splicing the cables together when they
met. The first message sent from Britain to the United States was a
cable from Queen Victoria to President James Buchanan delighting in
"an additional link between the nations whose friendship is founded
on their common interest and reciprocal esteem." The Ulster-Scots
Pennsylvanian Buchanan replied in even more Anglospherist language: "May the Atlantic telegraph, under the blessing of Heaven,
prove to be a bond of perpetual peace and friendship between our
kindred nations, and an instrument destined by Divine Providence to
diffuse religion, civilization, liberty, and law throughout the world."

Now the months that had passed "between the order and the

execution" in Burke's time were reduced to minutes. The phrase
"the abolition of distance," which sounds so contemporary to our
ears, had come into vogue by the 1860s. In his poem "The Deep-Sea
Cables," Kipling noted how the world had been altered:

> They have wakened the timeless Things; they have
>   killed their father Time;
> Joining hands in the gloom, a league from the last of the sun.
> Hush! Men talk to-day o'er the waste of the ultimate slime,
> And a new Word runs between: whispering, "Let us be one!"

Those who wanted a politically united Anglosphere were not
slow to see the implications. In 1884, the leading Liberal politician,
W. E. Forster, published a tract called *Imperial Federation*, which
took as its starting point the communications revolution:

> The inventions of science have overcome the great difficulties of
> time and space which were thought to make separation almost a ne-
> cessity, and we now feel that we can look forward, not to the isolated
> independence of England's children, but to their being united to one
> another with the mother-country, in a permanent family union.

Throughout the late nineteenth century, politicians and commentators proposed schemes for confederation under varying
names. Francis de Labillière, a barrister and former New Zealand
colonist, called it "Federal Britain." J. A. Froude, a novelist and historian, called it "Oceana." The name that caught on, though, was
coined by Sir Charles Dilke, another Liberal MP, who wrote his influential tract, *Greater Britain*, in 1868. His was the tag taken up by
the Imperial Federation League, which established active branches
in Australia, Barbados, British Guiana, Canada, New Zealand, and
the United Kingdom.

Then, as now, there were intense debates about the limits of the Anglosphere. Most advocates of Greater Britain aspired to a reconciliation of the two great branches of the English-speaking world: the American Republic and the British Empire. The brilliant Anglo-Canadian historian Goldwin Smith dreamed of uniting "the whole English-speaking race throughout the world, including those millions of men speaking the English language in the United States, and parted from the rest only a century ago by a wretched quarrel."

The idea had influential supporters in the United States, too. In 1897, in an editorial celebrating Queen Victoria's Diamond Jubilee, the *New York Times* declared, "We are a part, and a great part, of the Greater Britain which seems so plainly destined to dominate the planet."

Among the advocates of Greater Britain were the peerless American industrialist Andrew Carnegie and the African adventurer Cecil Rhodes. Yet there was a catch. These men were writing at a time when most public intellectuals, of left and right, were in some measure influenced by the then-voguish concept of racial determinism.

Greater Britain was propounded in essentially biological terms. The Canadian journalist John Dougall, for example, saw the federation of the United States and the British Empire as a sort of Nordic-Saxon ethnic reunification: "To England the alliance is desirable, as the future of the race seems as much connected with America as with England; to the United States it is desirable, as the past of the race belongs inalienably to England."

In his paper "The Reunion of Britain and America," Carnegie had explicitly called for a shedding of the nonwhite colonies, so as to facilitate a "race alliance" between the core Anglosphere states.

Cecil Rhodes created perhaps the most prestigious international scholarship program in the world, bringing young men from the colonies, the United States, and Germany to Oxford. He saw the

Germans as part of the same Teutonic race as the Anglo-Saxons;
he would have been astonished to see scholars, the cream of their
respective education systems, coming to Oxford from Bermuda,
Kenya, India, and Zambia in his name.

The idea of an Anglosphere defined by race was a product of a
relatively short-lived intellectual fashion. Even in the late nineteenth
century, commentators should have known that Anglosphere values
had been developed in a multiethnic context, that they had been
transmitted through intellectual exchange rather than gene flow.
Still, at a time when migration was slow and society homogenous, it
was understandable enough to think of nations as being defined by
race. Now, following the mass population movements of the twentieth century, few people define national identity in that way.

The Greater Britain of Carnegie and Rhodes had little appeal
to the growing number of Americans with non-British ancestry. As
well as a large African population, the United States had, by annexation, acquired a number of Mexican citizens, and the migration that had at first been overwhelmingly from the British Isles and
northwestern Europe was becoming wider in its scope.

But the biggest problem with the idea of the Anglosphere as an
extended Anglo-Saxon nation was that it made no room for India.
Some supporters of Greater Britain openly averred that Indians
could never fit in because, as Clive had claimed, they were by nature
autocratic, decadent, and pleasure-loving. Whereas the settler colonies had distilled and concentrated Britain's Whig-democratic culture, India was thought fit only for authoritarian rule. For which
reason, as the historian J. R. Seeley put it in *The Expansion of England*, "when we enquire then into the Greater Britain of the future,
we ought to think much more of our colonies than of our Indian
Empire."

The story of India since 1947 would have flabbergasted Seeley and his supporters. Despite poverty, ethnic and religious tensions, secessionist campaigns, and foreign wars, India has endured as a successful parliamentary democracy. Its legal system remains open to individuals seeking redress rather than being simply an instrument of the government—a characteristic that, more than any other, places India in the Anglosphere. Its economy becomes freer and more open by the day.

The British left India one particularly valuable legacy: unlike the former colonies of other empires, India began with a more or less functioning market economy in which property rights were respected and disputes arbitrated by an independent magistracy.

Moving from a measure of representative government to full democracy, however, was the achievement of an independent India. Like most peoples, Indians prospered under their own institutions. British administrators had been sincere in wanting to protect the Indian masses from the depredations of local despots. Yet there were limits to what foreign officials, however disinterested their motives, could do for such a vast country. Despite the best attempts of the Indian Civil Service, there were twelve famines during the Raj, the last in 1943. Since independence, there has not been one.

India itself, along with the successful Indian communities of the Anglosphere, constitutes teeming, noisy, polychromatic proof that Anglosphere values are transferable. When supporters of imperial unity balked at extending British rights to Indians, they told themselves that the ryots—the mass of Indian small farmers—were temperamentally unsuited to self-government; but perhaps, deep down, they feared the precise opposite. If Greater Britain lived up to its democratic ideals, then, as the Indian author Nirad Chaudhuri pointed out, its center of government must migrate to India, where two-thirds of its population lived. It was easier all round to avoid the question.

Nowadays, that conundrum has been left redundant by events. The idea of an Anglosphere as a monolithic bloc run by politicians is not only contrary to the spirit of the age but also contrary to the values that distinguish English-speaking civilization from its rivals. Anglosphere culture is based on self-government, localism, and the elevation of the individual over the state.

Modern Anglospherists want no truck with common currencies or federal parliaments or, indeed, any other joint state institutions beyond a flexible military alliance, a tariff-free zone, and perhaps a measure of free movement of labor. Their aim is a union of peoples, not of governments.

Throughout the nineteenth century, debates raged between those who sought to hasten the democratic independence of the colonies and those who wanted to draw them into some form of federation. Neither side exactly prevailed.

In 1901, one commentator, Bernard Holland, wrote that his ideal was "not a federal state, and not a mere alliance, but a thing between the two." That, in the twentieth century, is more or less what happened and, as a result, freedom was kept alive.

What next? Can the alliance of English-speaking peoples mutate and grow? Might we have a third Anglosphere century? To answer the question, let us break down the ingredients that underpinned its success, and recall the extraordinary alchemy that brought them together.

CHAPTER 8

From Empire to Anglosphere

I am in a country which scarcely resembles the rest of Europe. England is passionately fond of liberty, and every individual is independent.

—Baron de Montesquieu, 1729

The poorest man may in his cottage bid defiance to all the forces of the Crown. It may be frail, its roof may shake, the wind may blow through it, the storm may enter, the rain may enter; but the King of England cannot enter; all his force dares not cross the threshold of the ruined tenement!

—William Pitt the Elder, 1763

Toward the end of his life, Stanley Baldwin (1867–1947), three times the prime minister of the United Kingdom, was asked whether the thoughts of any political philosopher had guided him. Somewhat surprisingly—for Baldwin had never had much time for doctrines of any sort—he replied that, as a young man, he had been deeply impressed by the ideas of Sir Henry Maine, the eminent jurist and historian whose *Ancient Law* (1861) was a classic statement of Anglosphere exceptionalism. It was from that great text, Baldwin declared, that he had come to see that all human progress took the form of a movement from status to contract. Then, frowning, he paused. "Or was it the other way around?"

It is a telling story, an illustration of how even the most brilliant ideas can become stale through repetition. In Baldwin's day, Maine's status-to-contract theory was passing from seminal insight to orthodox theory. Nowadays it is barely read at all—except, it seems, by Francis Fukuyama, who drew heavily on it when formulating his own model of history as the inexorable triumph of liberal democracy.

Stop, though, and consider the sheer wonder of Maine's intuition. In almost every period of human history, people's circumstances were fixed at birth. In societies with negligible economic growth, people thought of assets as being fixed in quantity. Land was the only sure source of income, and those lucky enough to have it made sure that the system was rigged so that their children would enjoy the same advantage. Almost every social structure, from Neolithic times onward, had a caste aspect. Inca warrior priests, Indian Brahmins, ancien régime aristocrats, Soviet apparatchiks—all were beneficiaries of the closed and partially hereditary system that is the usual form of human organization. Slavery was almost ubiquitous.

The miracles of the past three and a half centuries—the unprecedented improvements in democracy, in longevity, in freedom, in literacy, in calorie intake, in infant survival rates, in height, in equality of opportunity—came about largely because of the individualist market system developed in the Anglosphere.

All these miracles followed from the recognition of people as free individuals, equal before the law, and able to make agreements one with another for mutual benefit.

In the twentieth century, German sociologists elaborated Maine's thesis, positing a shift from *Gemeinschaft* (community) to *Gesellschaft* (society). The essential difference between the two states lay in the freedom of the individual to make limited, case-by-case bargains with his fellows, rather than having to accept relationships defined by blood, ritual, or precedent. As the rationalist philosopher Ernest Gellner put it, "It is this which makes Civil Society: the forging of links which are effective even though they are flexible, specific, instrumental."

The dry language of social science disguises the vastness of the concept. Every farming society had elevated predation over production. Seizing someone else's crop offered a better effort-to-reward ratio than cultivating your own. Enshrining such predation in law, through tithes, taxes, and dues, was the most rewarding of all. Agrarian society gravitated toward oligarchy; most cultures have remained stuck there ever since.

In only one place was the pattern broken. Gellner, who saw the Anglosphere with the clear vision of an immigrant, wondered at "the circuitous and near-miraculous routes by which agrarian mankind has, *only once*, hit on this path" (emphasis in original).

It happened in the English-speaking lands from the late seventeenth century onward—though something similar was happening in contemporary Holland, which, but for an accident of geography and the expansionist ambitions of Louis XIV, might have been the place where the transformation was completed.

The path, once taken, led on to almost everything that we consider to be modern, comfortable, and rational, from human rights to the consumer society, from regular elections to equality for women. Pause, though, to survey the world, and see how exceptional these things are even today.

According to the Democracy Index, a survey of 167 states and territories published annually by the Economist Intelligence Unit, only 11.3 percent of people live in full democracies, and that population is overwhelmingly concentrated in the Anglosphere and the closely related and largely Protestant states of Nordic and Germanic

Europe. Strip out these two categories and the number of full democracies falls to seven: Czech Republic, Uruguay, Mauritius, South Korea, Costa Rica, Japan, Spain.

What catalyzed the shift from status to contract? What were the magic ingredients? In recounting our story so far, we have identified them. They are five in number.

First, the development of a nation-state: that is, a regime able to apply laws more or less uniformly to a population bound together by a sense of shared identity.

Second, and related to the fact of common nationality, a strong civic society: a proliferation of clubs, societies, and other groups filling the space between individual and state.

Third, island geography. The English-speaking world is an extended archipelago. With the exceptions of North America and India, its territories are insular: the Caribbean states, Ireland, Australia, New Zealand, the Falklands, Hong Kong, Singapore, Bermuda, Great Britain itself.

The Anglosphere is an essentially maritime civilization. "It has always seemed absurd to me that islands should not be English—unnatural," says Patrick O'Brien's fictional nineteenth-century naval captain, Jack Aubrey, when planning an attack on Mauritius.

Although geographically continental, the North American lands were politically isolated, "kindly separated by nature and a wide ocean," as Jefferson put it in his 1801 inaugural address, "from the exterminating havoc [of Europe]."

Fourth, religious pluralism in a Protestant context, which not only encouraged a proliferation of denominations but also encouraged an individualistic and democratic ethos—one that has long outlived its religious origins.

Fifth, and most important, the common law: a unique legal system that made the state subject to the people rather than the reverse.

If our thesis is correct, if these attributes were peculiar to the Anglosphere at the time of its takeoff, then we should expect foreign visitors during that period to remark upon them.

There were several European voyagers to Britain and America during the eighteenth and nineteenth centuries. We have enough evidence from travel journals and letters to build up a broad picture of what struck them as unusual. Several themes crop up repeatedly. Visitors found the English and American people undeferential, quarrelsome, keen on making money, fiercely individualistic, and uninterested in the doings of foreigners.

These attributes, of course, were by-products of the Anglosphere's political institutions, not the institutions themselves. The number of foreigners who were interested enough to write about the political and legal systems was smaller, but they included some of the most distinguished men of letters of the age, among them Voltaire, Montesquieu, and Tocqueville.

Throughout this period, visitors tended to see the United States as part of a broader "English" or "Anglo-Saxon" civilization. These days, few conservative seminars in the United States are complete without someone quoting Tocqueville in support of American exceptionalism. But the French aristocrat was as much a student of British as of American society. He traveled in both countries and, indeed, married an Englishwoman. Both places were the subject of his anthropological surveys.

He believed the United Kingdom and the United States formed a cultural continuum: "I do not think the intervening ocean really separates America from Europe. The people of the United States are that portion of the English people whose fate it is to explore the forests of the New World."

The reason Tocqueville remains popular is that he was an astute observer. Many of the distinguishing traits that he found in the United States—the commercialism, the individualism, the pluralism, the localism—continue to set that country apart nearly two centuries later. But Tocqueville did not believe that these characteristics had been acquired in the New World. For him, America's key advantage lay in what he called its "point de départ": the place it had started. English society had been characterized by representative assemblies, by resistance to taxation and state authority, by strong property rights. These tendencies, Tocqueville believed, had been given free rein in the Americas: "In the United States, the English anti-centralization system was carried to an extreme. Parishes became independent municipalities, almost democratic republics. The republican element which forms, so to speak, the foundation of the English constitution shows itself and develops without hindrance."

In other words, the American colonists had not just *borrowed* British political values and institutions; they had *intensified* them. Tocqueville believed that, just as French America had exaggerated the authoritarianism and seigneurialism of Louis XIV's France, and Spanish America the ramshackle corruption of Philip IV's Spain, so English America (as he called it) had exaggerated the libertarian character of the mother civilization.

Part of this heritage involved the proliferation of nonstate actors and civil associations: everything from privately funded orphanages to village bands. Tocqueville, like almost every foreign visitor, was struck by the way in which Anglosphere peoples would go ahead and form a club without seeking any kind of state license. "The spirit of individuality is the basis of the English character. Association is a means of achieving things unattainable by isolated effort. . . . What better example of association than the union of individuals who form the club, or almost any civil or political association or corporation?"

Such private associations—Edmund Burke's "little platoons"— were products of a state that was both strong and weak. Strong in the sense that it could be confident in the patriotism of its citizens (private clubs are never allowed to proliferate in insecure dictatorships); weak in the sense that it did not aspire to do by legislation what could easily be left to commercial or philanthropic endeavor.

But why did the nation-state develop so early, first in England, and then in the broader Anglosphere imperium? What lay behind the restraint of central government, and the flowering of the private associations? Again, foreign visitors were far more struck by the answer than natives: you couldn't reach the Anglosphere without crossing water.

THE ANGLOSPHERE ARCHIPELAGO

Insularity facilitated the development of a nation-state, as we saw in chapter 2. Critically, it also meant that there was no need for a permanent army. National defense was left largely to the navy, whose task was to prevent invaders from reaching the island in the first place. Other than in times of war, land forces were tiny, and largely made up of part-time territorial units.

Since neither a navy nor a territorial militia could easily be used for internal repression, the government found itself in a weak position vis-à-vis the population. When it wanted to pass a law, it had to secure the consent of the people through their representatives. When it needed revenue, it had to ask Parliament nicely.

Montesquieu was something of a geographical determinist, especially when it came to the model of British freedom that he admired:

> The inhabitants of islands have a higher relish for liberty than those of the continent. Islands are commonly of a small extent; one part of the people cannot be so easily employed to oppress the other; the sea separates them from great empires; tyranny cannot

so well support itself within a small compass; conquerors are stopped by the sea; and the islanders, being without the reach of their arms, more easily preserve their own laws.

While England had precociously become a nation-state in the ninth century, it was not quite an island state, for it shared the island of Great Britain with another kingdom. It is true that England was overwhelmingly the dominant partner in terms of wealth and population, sometimes treating Scotland as a semiprotectorate, occasionally receiving the ambiguous homage of Scottish kings. Nonetheless, the two crowns were not merged until 1603, nor the two parliaments until 1707.

It is no coincidence that the Anglosphere miracle followed the removal of Great Britain's last internal land border. Certainly Adam Smith, the greatest of all Scottish philosophers, saw the link. In one of his lectures at Glasgow University in 1763, he made the connection between isolation and liberty:

> The absolute power of the sovereigns has continu'd ever since its establishment in France, Spain, etc. In England alone, a different government has been established from the naturall course of things. The situation and circumstances of England have been altogether different. It was united at length with Scotland. The dominions were then entirely surrounded by the sea, which was on all hands a boundary from its neighbours. No foreign invasion was therefore much to be dreaded. We see that (excepting some troops brought over in rebellions and very impoliticly as a defence to the kingdom) there has been no foreign invasion since the time of Henry 3d.

Smith, like most educated Scots at that time, used "England" both in its narrow sense and as a synonym for Great Britain. His key insight, though, has to do with the redundancy of standing armies in island states:

> The Scots frequently made incursions upon them, and had they still continued seperate it is probable the English would never have recovered their liberty. The Union however put them out of the danger of invasions. They were therefore under no necessity of keeping up a standing army; they did not see any use or necessity for it. In other countries, as the feudall militia and that of a regular one which followd it wore out, they were under a necessity of establishing a standing army for their defence against their neighbours.

In Poland, France, and Sweden, argued Smith, kings were able to crush their legislatures: "The standing armies in use in those countries put it into the power of the king to over rule the Senate, Diet, or other supreme or highest court of the nation." England, and later the united Anglosphere, were different: "As the sovereign had no standing army he was obliged to call his Parliament."

Once again, the political tendencies of English-speakers in the Old World were exaggerated by those in the New. Americans— and, for that matter, Australians and New Zealanders—were even more hostile to standing armies, and even more anxious lest their leaders, under the pretext of fighting foreign wars, acquire powers that might facilitate despotism at home. The English had treated their liberties as an inherited tradition. The Americans were determined to take no risks, and wrote into their earliest charters guarantees against standing armies and in favor of private citizens' right to bear arms.

The United States might not literally be an island, but there is a reason why the attitude of nonintervention lauded from the early days of the republic is called "isolationism." The Founders, and those who have upheld their vision since, linked America's foreign policy to its geography. Theirs was the mentality of an island race.

Listen, for example, to George Washington's Farewell Address, still reverentially read out every year in the Senate:

> Our detached and distant situation invites and enables us to pursue
> a different course. . . . Why forgo the advantages of so peculiar a
> situation? Why quit our own to stand upon foreign ground? Why,
> by interweaving our destiny with that of any part of Europe, en-
> tangle our peace and prosperity in the toils of European ambition,
> rivalship, interest, humor or caprice?

As Tocqueville put it: "Placed in the middle of a huge continent with limitless room for the expansion of human endeavor, the
[American] Union is almost as isolated from the world as if it were
surrounded on all sides by the ocean."

Anglosphere political theory at the time posited a link between
geographical isolation, accountable government, trade, and peace.
Many of today's political scientists make a similar connection.
A comprehensive survey of the evidence by Henry Srebrnik at—
appropriately enough—Prince Edward Island University concludes:
"A number of studies suggest that island states are more likely to
be democratic than others, regardless of levels of economic development. The Commonwealth islands, especially, have done very well
on indices of political and civil rights and have provided the basis for
vibrant civil societies."

Island status, of course, is not the whole story. There are other
reasons that Guam is not Timor-Lest, that Bermuda is not Haiti,
that the Falklands are not the Comoros. Once again, these reasons
have to do with the *point de départ*, the starting point.

PROTESTANT ETHIC

There is a scene in *The Good Soldier*, Ford Madox Ford's exquisitely tragic 1915 novel, when the main characters go as tourists to see an original draft of the Protest: Martin Luther's 1517 denunciation of the abuses of the Roman Church. A wealthy American woman declares, superciliously:

> Don't you know that this is why we were all called Protestants? That is the pencil draft of the Protest they drew up. It's because of that piece of paper that you're honest, sober, industrious, provident, and clean-lived. If it weren't for that piece of paper you'd be like the Irish or the Italians or the Poles, but particularly the Irish.

That checklist of Protestant virtues could have been rattled off by English-speakers at almost any moment in the previous three hundred years. England had defined itself as the champion of the anti-Roman cause in Europe. Early America took this self-definition further, seeing itself as a providential nation, set aside by God through its religious orientation. Protestantism was the single biggest factor in the forging of a common British nationality out of the older English, Scottish, and Welsh identities—a common nationality then transmitted to the settler colonies.

Not until the twentieth century, though, did anyone attempt to test the idea of a Protestant ethic scientifically. In a series of essays published over 1904 and 1905, the German sociologist Max Weber asked whether there was a connection between Protestantism and economic growth. The resulting thesis, titled *The Protestant Ethic and the Spirit of Capitalism*, argued that there was indeed such a correlation, and that its cause lay in the peculiarly Protestant idea that industry and thrift were godly virtues. Until then, Weber argued, Christians had sought to renounce worldly things: to be ascetic, to despise wealth, to overcome desire. But Puritans believed that money, honestly acquired through hard work, was a sign of divine favor.

Plenty of critics rushed to point out flaws in Weber's methodology. Capitalism had not been an exclusively Protestant invention. Elements of it had developed in the city-states of northern Italy before the Dutch and the English perfected it. The Catholicism of the School of Salamanca is as libertarian as anything in the Protestant world. And Weber's thesis left little space for the equally powerful work ethic that we find in Austria, Bavaria, the Czech Republic, and the Catholic cantons of Switzerland. Indeed, for all the sneering of Ford Madox Ford's character, Ireland was a brilliantly successful example of tax-cutting capitalism until it made the disastrous decision to join the euro.

Still, Weber's findings cannot be easily dismissed. A major survey of the economic data between the years 1500 and 2000 carried out by Stanford University showed that, factoring for other variables, Protestant states began spectacularly to outgrow Catholic ones from the late seventeenth century, with "no signs of convergence until the 1960s." In 1940, GDP per capita in Europe's Protestant states was 40 percent higher than in its Catholic ones, and the divergence in the Americas was wider still.

How are we to explain such a huge gap? Weber's thesis of a different work ethic is, by its nature, difficult to measure. There are, though, other related explanations. For one thing, Protestantism, being Bible-based, placed a unique emphasis on literacy. Self-improvement and self-education were natural corollaries of a faith that encouraged worshippers to study the scriptures.

The spread in literacy might initially have been encouraged for devotional reasons, but it soon became an end in itself and, in northwestern Europe, the magical universe of the Middle Ages was gradually replaced with a rational one.

Protestant countries were not just keener on schools; they were keener on nonreligious schools. Indeed, they were keener on secularism in general. Again, it is difficult to separate out how much of this had to do with doctrine—a belief that the individual Christian should not rely on priests to intermediate his relationship with his Maker—and how much to do with the practical reality that, once the monopoly of a single church had been broken, it was impossible to prevent a general profusion of sects. As Voltaire put it, "If you have two religions in a land, they will be at each other's throats; but if you have thirty, they will dwell in peace."

Even in Protestant states where a single church was established, there was a logical impetus toward, first, toleration and, later, full equality, including the freedom to proselytize. Toleration, in its narrow sense, had existed in many multicultural states, including some which were in no sense liberal: Ottoman Turkey, for example. But total religious freedom was almost unknown outside the Anglosphere. Indeed, even within the Anglosphere, equality was slow to come to Catholics.

Nonetheless, visitors from Catholic Europe were surprised and delighted to find a place where you could be a free thinker without being anticlerical, and where no one was persecuted for his beliefs. Voltaire, who had fled to England in 1726 after falling out with a powerful aristocrat, became enraptured with this aspect of his new home. On one occasion, he was confronted by an angry London mob, who took him for a French spy. He pleaded with them that it was his immense misfortune not to have been born in Britain, and spoke with such conviction that the cheering crowd ended up carrying him back to his club on their shoulders. "By G— I do love the Ingles," he later wrote, in a delicious attempt to mimic the vernacular speech of his new friends. "G-d dammee, if I don't love them better than the French, by G—."

Montesquieu perceived, accurately, that granting equal status to different religions was a form of secularism, regardless of whether one sect remained established, and that it would lead to a proliferation of different churches:

> With regard to religion, as in this state every subject has free will, and must consequently be led either by the light of his own mind or by the caprice of whim, it follows that everyone will either look upon religion with indifference—which means that they will drift towards the established religion—or they must be zealous, by which means the number of sects is increased.

The Founders of the United States, yet again, pushed to its logical conclusion the tendency that had grown up in the rest of the Anglosphere, decreeing from the first that no religion should be established. Partly as a consequence, religious observance was and remains far more widespread in that country than in Europe. Nationalization is rarely successful: it stifles innovation and rewards inefficiency. Where state churches have dwindled, private—and privately funded—churches have fared better.

Full religious freedom—the removal of the minor remaining civil disabilities suffered by Catholics and Jews, such as not being able to become an MP—came to the United Kingdom nearly half a century after the establishment of the United States on the principle of religious pluralism. Still, to get a sense of how unusual this development was, it is worth remembering that the Anglosphere had embraced complete religious pluralism while the Spanish Inquisition was still operative: it was not finally wound up until 1834.

The centrality of Protestantism to the Anglosphere's cultural and political identity was the single biggest surprise to me when I researched this book. It is hard to recapture its importance today, partly because of the general decline of religion, and partly because the lines drawn by the Reformation have become blurred.

The Catholic Church moved briskly to address the outright abuses identified by the early Reformers. And, in an age when practicing Christians of any sort are a minority, the churches have tended to converge. Catholicism now places rather more emphasis on the Bible than it did; Protestantism, arguably, more emphasis on the Eucharist. Yet, though the doctrinal differences are less meaningful than they were, we can still see the traces of the *political* culture engendered by Protestantism, above all in the peculiar Anglosphere emphasis on individualism.

Tocqueville, too, was struck by the connection, and could see that that political culture now covered English-speakers of all faiths. Again, he traced it back to its *point de départ*: "When I consider all that has resulted from this first fact, I think I can see the whole destiny of America contained in the first Puritan who landed on these shores."

That Puritan—like those of his coreligionists who remained behind—had necessarily created a distinction between the public and private spheres, between state and church, between Caesar's realm and God's. Such a separation was not necessarily his intent— Massachusetts Congregationalists were almost the last people to support full equality for other Protestant sects, let alone for Catholics. But, whether he intended it or not, he created a political system where religious pluralism became inevitable.

To put it another way, the distinction was not between Catholic and Protestant *individuals*, but between Catholic and Protestant *states*. Tocqueville would have expected an Australian Catholic to be every bit as libertarian as his Protestant neighbors, a French Protestant every bit as committed to a strong state as his Catholic compatriots. "In fact," he remarked, "I never met an English Catholic who did not value, as much as any Protestant, the free institutions of his country."

This perhaps explains, as Weber's theory does not, the political culture of Bavaria, the Catholic parts of Switzerland, and, above all,

Ireland. What mattered was not people's view of the Immaculate Conception or priestly celibacy. What mattered was their view of personal liberty, free trade, and the inviolability of private contract. While Protestantism might have been an important component in *establishing* the Anglosphere's political culture, that political culture quickly took on a durability and energy that allowed it to flourish from Ireland to Singapore.

Even so, as we look back from our present, secular vantage point, we should nod respectfully at those who, as recently as the middle years of the twentieth century, acted on what they took to be the link between religion, industry and liberty.

Men like England's Alf Roberts, whose story might stand for those of millions. Alf left school at thirteen and, through sheer hard work, built a successful business as a grocer. A popular Methodist preacher, he saw commerce, faith, and politics as a continuum. "A lazy man," he declared in one of his sermons, "has lost his soul already."

Once a week, Roberts would bake extra loaves of bread and send them with his daughters to those of his neighbors whom he knew to be in need, taking care to explain that he had prepared too many, and would otherwise have to throw them out, for he didn't want anyone's pride to be hurt.

Throughout the 1930s and 1940s, Roberts was involved with many of the civic associations that are such a distinguishing characteristic of the Anglosphere. The moment that he realized Hitler was wicked was when he learned that the Rotarians, of whom he was a keen supporter, had been suppressed in Nazi Germany. When the war came, he organized the town's subsidized canteens for the men involved with war work.

For Alderman Roberts, thrift, sobriety, and hard work were not simply Protestant virtues: they were active political principles. He was a devoted town councilor, working constantly to keep expenditure down and cut the local rates. He saw his community, especially the shopkeeping class, as the real heroes of Britain, struggling on despite the taxes and regulations thrown at them by remote elites.

He was right. As Matt Ridley showed in *The Rational Optimist*, small businessmen have been the drivers of progress through the centuries. Societies that laud martial valor, nobility, and faith tend to be less pleasant places to live than societies that value freedom, enterprise, and privacy. The petit bourgeoisie, whom Marx so despised, have contributed more to human happiness than any number of crusaders. And they have done so, in the main, unhonored, unthanked, and unnoticed.

Indeed, the only reason we have heard of Alf Roberts today is that he drummed his values into his daughter who went on to become the greatest British leader, and perhaps the most enthusiastic Anglospherist, of the late twentieth century: Margaret Thatcher.

ANCIENT LAW

In 2000, in his book *Les Cartes de la France*, Hubert Védrine, then foreign minister of France, enumerated a series of qualities that made a state, as he put it, "un-European." It is as comprehensive a summary as you could ask of Anglosphere exceptionalism: "Ultraliberal market economy, rejection of the state, non-republican individualism, belief in the 'indispensable' role of the United States, and concepts which are Anglophone, common-law and Protestant."

It was, with the exception of the anti-American dig, a list that a Continental politician might have drawn up in 1700, 1800, or 1900.

Travelers from overseas noticed the limitations on the government, and the profusion of nonstate actors. They remarked—often with contempt—on the elevation of trade and moneymaking. They were struck, too, by the lack of deference among even the poorest classes, their readiness to assert their rights as freeborn men. And, if they came from Catholic countries, they could not help but gawk at the way the proliferation of different sects had led to a formal (in the United States) or de facto (in Britain) separation of church and state.

One thing, though, struck them as especially curious. They were astonished—as we, habituated to it by the passing centuries, are not—by the miracle of the common law. In their countries, laws were drafted by the government and then applied to particular cases. But in the Anglosphere (except Scotland), laws emerged case by case, building upward from the people rather than being handed down by the regime.

As an English Master of the Rolls named William Brett put it in a late-nineteenth-century ruling:

> The common law consists of a number of principles, which are recognized as having existed during the whole time and course of its existence. The Judges cannot make new law by new decisions; they do not assume a power of that kind: they only endeavour to declare what the common law is and has been from the time when it first existed.

In other words, neither judges nor politicians could alter the law, which, rather, was a heritage, passed down just as surely as any family heirloom. It was part of the compact that bound the living to the dead and to the yet unborn. As the seventeenth-century judge Robert Atkyns put it, "We ourselves of the present age, chose our common law, and consented to the most ancient Acts of Parliament, for we lived in our ancestors 1,000 years ago, and those ancestors are still living in us."

Lawyers raised in the Roman law tradition had, as they often still have, difficulty with this idea. Yet they could hardly miss the practical consequences of a legal system under popular control.

Independent magistrates, juries, the right of habeas corpus: all were remarked upon by visitors.

Montesquieu saw the common law as a glorious survival of the free, Germanic legal system, which, elsewhere, power-hungry kings had replaced with Roman law. England, believed Montesquieu, had been saved from such a fate by its island status, and had then passed on its unique legal system to its colonies.

Tocqueville, as usual, believed that the Americans had distilled their English heritage into a stronger and purer form. Nowhere else, he wrote, was the law so independent of the executive and legislative branches of the state, nor so accountable to the people expected to obey it. Along with the jury system—at which, like most observers, he marveled—Tocqueville was captivated by the idea that, instead of sending out centrally appointed judges and prefects, the Americans had localized their magistracy: "The Americans have borrowed from their English forefathers the conception of an institution which has no analogy with anything we know on the Continent—that of Justices of the Peace."

Adam Smith, whose native Scotland had, slightly incongruously, adopted Roman law, was struck by the same thing:

> One security for liberty is that all judges hold their offices for life and are intirely independent of the king. Everyone therefore is tried by a free and independent judge. The Habeas Corpus Act is also a great security against oppression, as by it any one can procure triall at Westminster within 40 days who can afford to transport himself thither. Before this Act the Privy Councill could put any one they pleased into prison and detain him at pleasure without bringing him to triall.

Again and again in our story, we have seen the common law used as a bulwark against excessive state power. It served heroically against Charles I and again against James II; it found that the air of England was so pure that no man who breathed it might remain a slave; it made the American Revolution.

Above all, common law has proved the surest defense of property rights. Today, companies from all over the world pay premiums to sign their contracts in common-law jurisdictions. They do so because they have confidence in the impartiality, security, and fairness of the system.

Every year, the Heritage Foundation ranks the economic freedom of the world's states, comparing such data as corporate tax rates, personal tax rates, security of ownership, and how long it takes to found a business. In 2012, six of the top ten territories were Anglosphere states. The top four, in order, were Hong Kong, Singapore, Australia, and New Zealand. In no age but our own would it be considered impolite or impolitic to point out what they had in common.

NO EXTREMISM, PLEASE

One Anglosphere leader who was never bothered about being impolitic was Alderman Roberts's younger daughter. On one occasion, Margaret Thatcher scandalized *bien-pensant* opinion by observing, tartly but truthfully, that, throughout her lifetime, Britain's problems had come from Europe, and the solutions to those problems from the English-speaking world.

She thought she knew why. Although she was no intellectual—she would have had strong words for anyone who suggested otherwise—she nevertheless had a sense that Anglosphere exceptionalism was rooted in history.

In 1989, she was invited to Paris, along with other world leaders, to celebrate the bicentenary of the French Revolution. She felt, in her bones, that it was wrong to go. It wasn't just that the French Revolution had ushered in a series of wars with Britain, which lasted almost uninterruptedly for a quarter of a century. It was that the values of that revolution—the statism, the violence, the belief in enforced equality, the anticlericalism—were the opposite of everything she took to be the true basis of freedom.

François Mitterrand, that wiliest of French presidents, decided to host the G7 summit in Paris over the date of the bicentenary, thereby more or less obliging the British prime minister to attend. She was not happy, and gave vent to her feelings in an interview with a French newspaper:

> Human rights did not begin with the French Revolution; they stem from a mixture of Judaism and Christianity. We had 1688, our quiet revolution, where Parliament exerted its will over the King. It was not the sort of Revolution that France's was. "Liberty, equality, fraternity"—they forgot obligations and duties I think. And then of course the fraternity went missing for a long time.

Again, the words were undiplomatic, but the underlying analysis was accurate. Anglosphere exceptionalism does not reside in democracy. *Democracy* is an old word for an old idea, and the Continental revolutionary tradition is at least as much based on majority rule as the Westminster parliamentary tradition.

The difference lies, rather, in the approach to the rule of law, property rights, and personal freedom. The post-Jacobin Continental strain of democracy elevated majority rule over individual liberty. Anglosphere democrats were a different breed altogether.

European radicalism grew up alongside, but largely separate from, Anglosphere radicalism. Its heritage and assumptions were only very distantly related to those of the Levelers, the Chartists, or the early English-speaking trade union activists. Its ultimate philosophical inspiration came rather from the collectivist writings of Hegel and Herder and, in particular, from Rousseau's belief in the "general will" of the people in place of the private rights of the citizen. Here was a philosophy that conceived of rights as universal, guaranteed by law and handed down by the government rather than inherited—a philosophy very different from the common-law conception of a free society as an aggregation of free individuals.

The Continental model had an obvious flaw: the contracting out of human rights to a charter, necessarily interpreted by some state-appointed tribunal, left the defense of freedom in a small number of hands. If those hands failed, freedom failed. In the Anglosphere, where the defense of freedom was everyone's business, dictatorship and revolution were almost unknown. Liberties, as the novelist Aldous Huxley observed, are not given, they are taken.

To make the point slightly differently, liberty was theoretical in Europe, practical in the Anglosphere. As the nineteenth-century Conservative prime minister Benjamin Disraeli put it, "To the liberalism they profess, I prefer the liberty we enjoy; to the Rights of Man, the rights of Englishmen."

From the first, the radical tradition in Europe was violent. The repression that followed the French Revolution is known as "the Terror." That name, however, was not bestowed by opponents of the Revolution; on the contrary, it was taken up by the Jacobins themselves. On September 5, 1792, the revolutionaries announced their policy in these terms: "It is time that equality bore its scythe above all heads. It is time to horrify all the conspirators. So, legislators, place Terror on the order of the day! The blade of the law should hover over all the guilty."

The Continental radical tradition eventually whelped two malformed pups: revolutionary socialism and fascism, respectively the most murderous and second-most murderous ideologies in human history. Hayek, another clear-eyed and appreciative immigrant to the Anglosphere, correctly understood that the two statist ideologies were related, both in their origins and in their elevation of the herd over the individual. Both, in their different ways, were antitheses to Anglosphere liberalism.

Between the two world wars, fascism and communism were seen as the coming ideas, and country after country fell for—or at least fell to—the one or the other. Antiparliamentary forces did not rely on coups and invasions: from Portugal to Russia, from Germany to Greece, autocratic government enjoyed broad popular support. In the Anglosphere, by contrast, no fascist party ever managed to get a single candidate elected to a national legislature, and the fortunes of the revolutionary left were almost as dire.

In the United States, socialism was seen—with some justice, given who its chief proponents were—as a creed for a minority of European immigrants. Even at its high point in the 1920s, it never had more than a hundred thousand active supporters.

In Australia and New Zealand, communist parties existed throughout the lifetime of the Soviet Union, and were spasmodically active in the trade union movement, but never managed to elect a single MP.

In Canada, following the entry of the Soviet Union into the Second World War, the former communists formed a broader party known as Labor-Progressive, which at its peak managed to elect one MP, Fred Rose—who was later found guilty of spying for the Soviets.

In Ireland, the communist party never had any success at the polls, though today's Socialist Party, a Trotskyist offshoot from Labour, does have a single parliamentary representative: the likable Euro-skeptic Joe Higgins.

In the whole history of the United Kingdom, a grand total of six revolutionary socialists have been elected to the House of Commons: two in 1922, one in 1935, two more in 1945, and one—the antiwar radical George Galloway—in 2005. Given that, throughout this period, the House of Commons had more than six hundred MPs, it's an extraordinarily thin record.

Of all the Anglosphere states, only India, and, more notably, South Africa, whose communist movement benefited from its opposition to apartheid and fought elections in alliance with the ANC, have seen any electoral success for the revolutionary extremes.

It is easy to forget the mass electoral support that fascist and communist parties secured across Europe, not just between the wars, but well into the second half of the twentieth century. Membership of the Communist Party of Great Britain peaked at 60,000. In France, there were 800,000 Communist Party members, in Italy 1.7 million.

Many creeds took root in the fertile loam of the Anglosphere, but the seeds of fascism and communism never sprouted there. The democratic constitutions that most European states had acquired after the First World War gave way one after another before the men in uniforms. But, in the Anglosphere, the defense of civil rights was everyone's duty. People took their responsibilities seriously, and extremists were rejected again and again at the polls.

Fascism never held much appeal for the English-speaking peoples. It seemed faintly ludicrous, with its parades and shiny boots and silly salutes. Nor did the mainstream Anglosphere left ever flirt seriously with Marxism.

Whether from historical experience or from some quality in their language, the English-speaking peoples have tended to prefer practice to theory, to shy away from abstract ideologies. And communism, for all the claims of its proponents, is the most abstract ideology of all.

Marxism, uniquely among political philosophies, defined itself as a science. To its adherents, its propositions were not speculative but empirical. As a good Hegelian, Marx saw his forecasts as part of an inexorable historical process. Yet every one—*every one*—of them turned out to be false.

Capitalism was supposed to destroy the middle class, leaving a tiny clique of oligarchs ruling over a vast proletariat. In fact, capitalism has enlarged the bourgeoisie wherever it has been practiced. Capitalism was supposed to lower living standards for the majority. In fact, the world is wealthier than would have been conceivable 150 years ago. The whole market system was supposed to be on its last legs when Marx and Engels were writing. In fact, it was entering a golden age, hugely benefiting the poorest. As the economist Joseph Schumpeter put it, the princess was always able to wear silk stockings, but it took capitalism to bring them within reach of the shop-girl. The living standard of someone on benefits in the Anglosphere today is higher than that of someone on average wages in the 1920s.

I don't know how many of the people parroting Marx are aware that they're doing so. But, whatever name we call it by, his doctrine has proved stunningly impervious to events. You'd have thought—I did think—that the collapse of the Warsaw Pact regimes in 1989 would have definitively refuted revolutionary socialism. Yet successive generations continue to fall for it.

The more we learn of behavioral psychology, the more we understand that ideologies are as much a product of people's nature as of observed experience. The perverted doctrines that actuated the Bolshevists may be immanent in a portion of humanity. Some people are determined to see every success as a swindling of someone else, every transaction as an exploitation, every exercise in freedom as a violation of some ideal plan, every tradition as a superstition. How delicious that, as we approach the bicentenary of his birth, Karl Marx should have turned into the thing he loathed above all: the prophet of an irrational faith.

English-speaking countries have been fortunate in the temper of their left-wing parties. You don't have to look far to find socialist movements rooted in envy, authoritarianism, and bloodthirstiness. Significant parts of the Continental left were born in violence and revolution—a revolution that would be complete, according to its agents, "only when the last king has been strangled with the entrails of the last priest."

In the United States, no socialist movement of any significance got off the ground at all, and both main parties were, in European terms, center-right. In the other main Anglosphere territories, the left tended to grow out of the radical wing of Whig-Liberalism.

From the first, the Anglosphere's labor parties were associated with self-help and self-improvement: with brass bands, the temperance movement, workingmen's libraries. Morgan Phillips, the Welsh colliery worker who served as the British Labour Party's secretary-general between 1944 and 1961, declared that that party owed "more to Methodism than to Marxism."

There was much truth in that remark. While there has always been an angry and sour element in Labour, it is balanced by a different tendency: one that seeks to improve the lot of the poor, not by tearing down the system, but by extending opportunities.

The proudest achievements of the Anglosphere left, down the years, have involved the dispersal of power from closed elites to the general population. This high-minded ambition led to legal rights for women, the extension of the franchise, and universal education. These reforms happened throughout the English-speaking world, though not always at the same moment: New Zealand, for example, became the first state to extend the vote to all adults in 1893, though it was followed swiftly enough by other Anglosphere nations. In no major English-speaking country, though, did any significant number of people argue that the changes they sought required the system to be overthrown. The twin settlements of 1689 and 1787 lasted: to call for a revolution in the English language sounded affected, foreign, or downright childish.

All of this raises an intriguing question. Why is patriotism, in English-speaking societies, mainly associated with the political right? After all, measured against almost any other civilizational model, the Anglosphere has been overwhelmingly progressive.

It is true that the individualism of English-speaking societies has an antisocialist bias: there has always been a measure of resistance to taxation, to state power, and, indeed, to collectivism of any kind. But look at the other side of the balance: equality before the law, regardless of sex or race; secularism; toleration for minorities; absence of censorship; social mobility; universal schooling. In how many other places are these things taken for granted?

Why, then, is the celebration of national identity a largely conservative pursuit in English-speaking societies? It won't do to say that patriotism is, by its nature, a right-of-center attitude. In the Continental tradition, if anything, the reverse was the case. European nationalists—those who believed that the borders of their states should correlate to ethnic or linguistic frontiers—were, more often than not, radicals. The 1848 revolutions in Europe were broadly leftist in inspiration. When the risings were put down, and the old monarchical-clerical order reestablished, the revolutionaries overwhelmingly fled to London, the one city that they knew would give them sanctuary. With the exception of Karl Marx, who never forgave the country that had sheltered him for failing to hold the revolution that he forecast, they admired Britain for its openness, tolerance, and freedom.

What stops so many English-speaking leftists from doing the same? Why, when they recall their history, do they focus, not on the extensions of the franchise or the war against slavery or the defeat of Nazism, but on the wicked imperialism of, first, the British and, later, the Americans?

The answer lies neither in politics nor in history, but in psychology. The more we learn about how the brain works, the more we discover that people's political opinions tend to be a rationalization of their instincts. We subconsciously pick the data that sustain our prejudices, and block out those that don't. We can generally spot this tendency in other people; we almost never acknowledge it in ourselves.

A neat illustration of the phenomenon is the debate over global warming. At first glance, it seems odd that climate change should divide commentators along left–right lines. Science, after all, depends on data, not on our attitudes to taxation or defense or the family.

The trouble is that that we all have assumptions, scientists as much as anyone else. Our ancestors learned, on the savannahs of Pleistocene Africa, to make sense of their surroundings by finding patterns, and this tendency is encoded deep in our DNA. It explains the phenomenon of cognitive dissonance. When presented with a new discovery, we automatically try to press it into our existing belief system; if it doesn't fit, we question the discovery before the belief system. Sometimes this habit leads us into error. But without it, we should hardly survive at all. As Edmund Burke argued, life would become impossible if we tried to think through every new situation from first principles, disregarding both our own experience and the accumulated wisdom of our people—if, in other words, we shed all prejudice

If you begin with the belief that wealthy countries became wealthy by exploiting poor ones, that state action does more good than harm, and that we could all afford to pay a bit more tax, you are likelier than not to accept a thesis that seems to demand government intervention, supranational technocracy, and global wealth redistribution.

If, on the other hand, you begin from the proposition that individuals know better than governments, that collectivism was a demonstrable failure, and that bureaucracies will always seek to expand their powers, you are likelier than not to believe that global warming is just the left's latest excuse for centralizing power.

Each side, convinced of its own bona fides, suspects the motives of the other, which is what makes the debate so vinegary.

Proponents of both points of view are quite sure that they are dealing in proven facts, and that their critics must therefore be either knaves or fools.

The two sides don't simply disagree about the *interpretation* of data; they disagree about the data. Never mind how to respond to changes in temperature; there isn't even agreement on the extent to which the planet is heating. Though we all like to think we are dealing with hard, pure, demonstrable statistics, we are much likelier to be fitting the statistics around our preferred worldview.

Central to the worldview of most people who self-identify as left-of-center is an honorable and high-minded impulse, namely support for the underdog. This impulse is by no means confined to leftists; but leftists exaggerate it, to the exclusion of rival impulses.

Jonathan Haidt is a psychologist, a man of the soft left, who set out to explain why political discourse was so bitter. In his seminal 2012 book, *The Righteous Mind*, he explains the way people of left and right fit their perceptions around their instinctive starting points. Support for the underdog, in conservatives, is balanced by other tendencies, such as respect for sanctity. In leftists, it is not.

Once you grasp this neurological difference, all the apparent inconsistencies and contradictions of the leftist outlook make sense. It explains why people can think that immigration and multiculturalism are a good thing in Western democracies, but a bad thing in, say, the Amazon rain forest. It explains how people can simultaneously demand equality between the sexes and quotas for women. It explains why Israel is seen as right when fighting the British but wrong when fighting the Palestinians.

History becomes a hierarchy of victimhood. The narrative is fitted around sympathy for downtrodden people. The same group can be either oppressors or oppressed depending on the context. Hispanic Americans, for example, are ranked between Anglos and Native Americans. When they were settling Mexico, they were the bad guys; when they were being annexed by the United States, they were the good guys.

All historians, of course, have their prejudices. My purpose is simply to explain why national pride in Anglo-American culture is so concentrated on one side of the political spectrum. The answer, quite simply, is that there are very few scenarios in which the Anglosphere peoples can be cast as the underdogs. Small countries take satisfaction in their struggles against mightier neighbors, and that pride is shared across the parties. In many former colonies, patriotism is seen as an essentially leftist cause, and conservatism is associated with, if not exactly collaboration, certainly cultural subservience to the former power. In the Ba'athist Arab states, in Sandinista Nicaragua, in Peronist Argentina or Bolivarian Venezuela, nationalism was a revolutionary socialist creed, associating popular sovereignty with state power, the toppling of unpatriotic oligarchies, and the removal of foreign influence.

Anti-American and anti-British agitators around the world have taken up nationalist language—the only nationalism of which English-speaking progressives generally approve. George Orwell wrote disparagingly of "the masochism of the English Left": its readiness to ally with any cause, however vile, provided it was sufficiently anti-British. He cited the IRA and Stalinism. Had he been writing today, he'd doubtless have extended the critique to the American left and Islamism.

Nationalism is fine for leftist opponents of the Anglosphere: Welsh nationalists, *anti-yanqui* agitators in Latin America, Quebec separatists. All are able to slot their sense of nationhood into the hierarchy of victimhood, to see themselves as underdogs. Anglosphere progressives, by contrast, can rarely do so. The few occasions when they can—Washington leading his exhausted men through the snow at Valley Forge, Churchill rallying London during the Blitz—are commensurately popular. But there is no getting away from the fact that the Anglosphere, over the past three hundred years, has generally been more technologically advanced than other civilizations.

This very success makes it awkward to celebrate the distinguishing features of Anglosphere culture. To do so is to risk the appearance of complacency or jingoism. Episodes that, in any other context, leftists would uncomplicatedly applaud—Napier's response to the Hindu priests who favored sati, for example—are tainted by their supposed cultural imperialism or colonial arrogance.

The tendency to what Orwell called masochism is the perversion of a healthy Anglosphere characteristic, namely fair-mindedness. We like to think, we English-speaking peoples, that we are tolerant, that we look at things from other people's point of view. It is not hard to see how this trait can be exaggerated to the point where it becomes, if not exactly self-hatred, certainly a form of cultural relativism in which the unique achievements of Anglosphere civilization are devalued.

The tendency has existed ever since the Anglosphere countries rose to global dominance. In their 1885 operetta *The Mikado*, Gilbert and Sullivan mocked

> *The idiot who praises, with enthusiastic tone,*
> *All centuries but this and every country but his own.*

Until the second half of the twentieth century, though, cultural relativism was largely confined to university campuses and small circles of intellectuals. Even those most determined to see the other point of view, to find fault in their own civilization, generally had to admit that the Anglosphere was a freer, fairer, and more progressive society than Stalinist Russia or Republican China or Imperial Ethiopia.

Still, ideas have consequences. The students of the relativist academics became schoolteachers. Their pupils ended up imbibing a version of the doctrine. It is now quite normal, in English-speaking societies, to approach our history with guilt rather than pride. It won't quite do to say that we have done away with moral judgment. But, absurdly, we judge the deeds of the English-speaking peoples according to contemporary leftist nostrums rather than by the standards of their times.

As we lose sight of what the Anglosphere has achieved, we risk losing the institutions that have served to make it what it is.

CHAPTER 9

Consider What Nation It Is
Whereof Ye Are

Lords and Commons of England, consider what Nation it is whereof ye
are, and whereof ye are the governors: a Nation not slow and dull, but of
quick, ingenious, and piercing spirit, acute to invent, suttle and sinewy
to discours, not beneath the reach of any point the highest that humane
capacity can soar to.

—JOHN MILTON, 1644

All nations change over time. We have wandered far for many years. Yet
our constitutional faith has not been erased from our consciousness. Nor
has it been defeated in our politics. Our principles always await rediscov-
ery, not because they are written on faded parchments in glass cases, but
because the immutable truths of liberty are etched on the human soul.

—MATTHEW SPALDING, 2009

In July 2012, the Republican presidential candidate, Mitt
Romney, went on an international tour designed to burnish his
foreign policy credentials. He chose three traditional allies that
had been neglected by the Obama administration: Israel, Poland,

and the United Kingdom. The Israeli and Polish legs went well enough, but the British visit was overshadowed by a remark supposedly made by one of his aides—though the claim was contested—before he set out. The United States and the United Kingdom, the staffer was quoted as saying, were bound together by their common Anglo-Saxon heritage.

Leftists on both sides of the Atlantic leaped on the remark, affecting to interpret it as a racial slur. The Obama campaign called it "stunningly offensive"—though, like most people claiming to be offended, they almost certainly knew better. After all, it must have been clear to even the most ignorant Romney staffer that Anglo-Saxons, as an ethnic category, are a small percentage of the U.S. electorate. As they are, for that matter, of the British electorate. It's not just that my country is full of people with non-Saxon surnames like Hannan. As we saw in chapter 1, the Anglo-Saxons intermarried with other peoples from the very beginning, and "Anglo-Saxon values" were developed in a multiracial context. Even those pretending to be the most upset surely knew that the reference was cultural rather than ethnic.

One of the great virtues of Anglo-Saxon political culture is precisely that it conceives nationhood in nonethnic terms. Americans take justified pride in their successful assimilation of newcomers. Millions have been drawn to their country from every scrap of dry land determined to become American.

What do we mean by "becoming American"? When we break it down, there are three irreducible elements. First, accepting the values encoded in the U.S. Constitution: free speech, the division of powers, religious toleration, and so on. Second, understanding the unwritten codes bound up with those values: civic engagement, open competition, private contract, equality for women. Third, speaking English. And where do these three characteristics have their origin? Not in Korea or Romania or Ecuador, though people from those places can adopt them as easily as anyone else. What we mean by Anglo-Saxon civilization is the set of cultural, social, and political assumptions parceled together in the English-speaking world.

The fact that the row damaged the Romney campaign as much as it did is a sad reflection on how we have come to look at our past. (I hope that, if you have read this far, you will by now appreciate that I am using "we" to mean the free English-speaking peoples.) Instead of taking Anglo-Saxon values to mean personal liberty, the common law, representative government, resistance to tyranny, jury trials, and all the rest, the staffer's remark was pressed into the currently dominant narrative where the people who spread those values find themselves at the wrong end of the victimhood hierarchy.

In every core Anglosphere country, we find that those most uncomfortable with that legacy tend also to be those who reject the notion of the Anglosphere as a global alliance. Those in Britain who insist that the Commonwealth is a relic of empire and that their country's destiny is in Europe have their precise equivalents—socially, politically, and culturally—among those Australians who say that the link with Britain is anachronistic, and that Australia must embrace its future as an Asian state.

Such sentiments are common enough among the academic and media elites but have little appeal to the general population. When Australia voted on whether to abolish the monarchy in 1999—a vote that was widely seen as having less to do with the institution of the monarchy than the country's relationship with the rest of the Anglosphere—republican leaders, who were generally of British or Irish descent, could barely hide their frustration with immigrants from elsewhere who preferred the political model they had found in Australia to that of the states they or their ancestors had left, and who voted for the status quo.

The Internet and the proliferation of online media have revealed the numerical weakness of the anti-Anglosphere elites, who once dominated the airwaves. Political leaders like Canada's Stephen Harper, Australia's Tony Abbott, and New Zealand's John Key take their alliance with one another, and with the United States and Britain, for granted. They know—because they can read opinion polls—that the onus is on their opponents to change minds.

Sympathy, in the literal sense of fellow-feeling, is a powerful force in human affairs. It affects many who are in no sense Anglosphere apostles. I recently found myself sitting at dinner next to Michael Ignatieff, the Canadian Liberal leader whom Stephen Harper defeated in 2011. A clever and decent man, he had had an enormously successful career as an academic and broadcaster, but proved too innocent to make a successful transition into politics. As we talked, he set out all the reasons why, as he saw it, the Anglosphere was not the force it was. Canada, he told me, had seen its British orientation diluted by immigration. His own name was Ignatieff, he reminded me—a touch unnecessarily, perhaps. All right, I replied, but whose side would you be on if there were a war between, oh I don't know, Australia and Indonesia—even if you believed that the Australians were the guiltier party? He paused. "Fair enough," he said at length.

For a long time, it was difficult to promote the Anglosphere without connotations of imperialism or dominance. In Ireland and India, especially, the focus of cultural and political development was differentiation from the former colonial power. Now, though, the independence struggles have almost faded from memory. It is possible to value the parliamentary common-law model of the Anglosphere without seeming to want the old system back.

The Internet is an Anglosphere invention and remains overwhelmingly an English-language tool. The Anglosphere peoples are drawn together by it, establishing cultural, commercial, and social links independent of government. James Bennett, who popularized the Anglosphere concept, believes that the English-speaking democracies are uniquely well placed to form what he calls "a

devolved, flexible, network commonwealth." Within that network
are particular cultural links. The massive Indian diaspora is drawn
together by films, music, and other cultural continuities that exist
overwhelmingly within an Anglosphere context.

When Sile de Valera, Éamon's granddaughter and then an Irish
cabinet minister, declared in 2000 that most Irish people felt more at
home in Boston than in Brussels, and that the EU had gone too far
in harmonizing the identities of its member states, she was stating an
obvious truth: there are Irish networks within the Anglosphere that
connect every core territory. Yet she was also marking a major change
in orientation. Ireland was no longer turning to the EU as a way of
underlining its independence from the United Kingdom. It was confident enough to embrace an Anglosphere identity as an equal partner.

People of Ulster Protestant origin have also found that the Internet facilitates the development of networks within the Anglosphere. The hillbillies of the Appalachians were so called because,
in memory of the victory of William III ("King Billy") at the Boyne,
they marched every year on July 12. The past decade has seen a dizzying expansion of the cultural links among Ulster-descended communities in North America, Australia, Great Britain, and Ireland.

Relations among the core Anglosphere states are better than they
have been since the Second World War. Canada, Australia, and New
Zealand base their foreign and defense policies around the alliance
with the United States, and, to a lesser degree, the United Kingdom.
The United Kingdom and Ireland are closer than at any time since
the partition. India has embraced an alliance with the United States
and, by implication, with the rest of the English-speaking world,
that suddenly seems the most natural thing on earth. When, in the
aftermath of the terrible tsunami of 2004, the U.S., Australian, and
Indian navies collaborated in the relief effort, they found that they
had a degree of interoperability that didn't exist even among NATO
allies after nearly six decades of joint operations.

INVENTING FREEDOM

Of the core Anglosphere governments, only one now stands purposefully aside: that of the United States.

There's no getting around it: Barack Obama doesn't like the Anglosphere and, in particular, doesn't like the British. Let's review the evidence. The president received from Gordon Brown a pen-holder made from the timbers of HMS *Gannet*, who had spent her life at sea prosecuting a successful war against the slave trade. He reciprocated with a set of DVDs.

In his first summit meeting with a British prime minister, he silkily downgraded the two countries' alliance. American presidents have to be careful with their superlatives, of course, lest they offend Mexico or Israel or some other favored nation, yet previous U.S. leaders had no difficulty describing the United Kingdom as "our closest ally." George W. Bush cheerfully went further, calling the United Kingdom "our closest friend and strongest ally," and adding, on the day Baghdad fell in 2003: "America has no finer ally than Great Britain." For the forty-fourth president, however, Britain was simply "one of our allies," right up there with Bahrain and Honduras.

When an oil slick caused millions of dollars of damage in Louisiana and other parts of the Gulf of Mexico area, President Obama repeatedly and calculatedly used the occasion to attack an imaginary company called "British Petroleum." In fact, no such corporation had existed for a decade: it had become "BP" following a merger with Amoco that gave it as many U.S. as British shareholders.

When George W. Bush met the queen, he made a characteristic verbal slip and then, with an unrehearsed grin, remarked: "She gave me a look like only a mother could give a child." It was a touching phrase, a gracious acknowledgment of America's British heritage. His successor? His successor presented Her Majesty with an iPod loaded with . . . his own speeches.

Visiting West Africa for the first time, President Obama made a point of referring to the region's struggle for independence from

Britain (not that it was much of a struggle: most African colonies were brought to self-government without a shot being fired in anger). Yet, despite touring the slave stations, he managed to avoid any mention of the fact that the slave trade had been extirpated largely by the Royal Navy.

His coolness goes beyond symbolism or etiquette. The present administration has come dangerously close to pressing the position of Peronist Argentina on the Falkland Islands. In a series of resolutions, it has lined up with such virulently anti-Western figures as Venezuela's late Hugo Chavez and Nicaragua's Daniel Ortega to back Argentina's demands for negotiations on sovereignty. The State Department has even taken to referring to the disputed territories by their Argentine name, Las Malvinas.

Obama has every right to pick his friends, of course. What is significant here is the link between his dislike of the United Kingdom and his rejection of the values and institutions that the early Americans took from the British Isles. Obama rejects not just the special relationship with Britain, but the worldview on which it rests.

The conventional explanation for the president's Anglophobia is that he is bitter about the way his grandfather, Hussein Onyango Obama, was interned during the Mau Mau rebellion, a partly tribal war in Kenya in the 1950s between pro-British and pro-independence factions. But this explanation doesn't fit with what Obama himself has written.

Barack never knew his grandfather, but what he later found out repelled him. Despite being detained by the British authorities, Onyango had remained something of an imperialist, believing that the British had earned their place in Kenya through superior organization. He even used to argue that Africans were too lazy to make a success of independence. The young Obama was horrified: "I had imagined him to be a man of his people, opposed to white rule," he wrote in *Dreams from My Father*. "What Granny [Sarah Obama, one of Onyango's wives] had told me scrambled that image completely, causing ugly words to flash across my mind. Uncle Tom. Collaborator. House nigger."

No, the president's antipathy comes not from the grandfather he disdained, but from the father he worshipped—albeit from a distance. Barack Obama Sr. abandoned Obama's mother and had almost nothing to do with the young Barry (as he was known throughout his childhood and adolescence). He did, however, make one journey to Hawaii that had an enormous impact on the ten-year-old future president. Barry, as boys sometimes do, had been telling tall tales about his absent father. He had implied to his classmates that Barack Sr. was a great chief, and that he would himself one day inherit the tribal leadership. He was mortified when his class teacher asked his father to talk to the class; Barry was afraid that his fibs would be exposed. His anxieties vanished as the handsome Kenyan strode into the room in African dress and proceeded to give a talk that was the defining moment of Barry's childhood:

> He was leaning against Miss Hefty's thick oak desk and describing the deep gash in the earth where mankind had first appeared. He spoke of the wild animals that still roamed the plains, the tribes that still required a young boy to kill a lion to prove his manhood. He spoke of the customs of the Luo, how elders received the utmost respect and made laws for all to follow under great-trunked trees. And he told us of Kenya's struggle to be free, how the British had wanted to stay and unjustly rule the people, just as they had in America; how many had been enslaved only because of the color of their skin, just as they had in America, but that Kenyans, like all of us in the room, longed to be free and develop themselves through hard work and sacrifice.

That view of a constant struggle by the oppressed against the oppressors, the colonies against the colonizers, the have-nots against the haves, became the mainstay of Barack Jr.'s worldview. In *The Roots of Obama's Rage*, Dinesh D'Souza showed beyond doubt that Obama's domestic and foreign policy agendas were inspired by his father's 1950s anticolonialism. The mistake that every other analyst has made, argued D'Souza, was to try to fit Obama into America's racial narrative. But the battle for civil rights was only tangentially a part of his story. Indeed, he infuriated many black political organizations by refusing to take up the issues that they cared about, such as the minimum wage and affirmative action. His struggle was not that against segregation in Mississippi but that of African protectorates against European empires.

For example, Obama's climate change policies make little sense either as an attempt to slow global warming or as a way to make the United States more popular. But they make perfect sense as a mechanism for the redistribution of wealth from rich nations to poor. (D'Souza noted, as an instance, the way in which the Obama administration banned offshore drilling in the United States while sponsoring it in Brazil.) The same is true of his enthusiasm for nuclear disarmament. It seems bizarre to be pursuing the elimination of atomic weapons in a forum that doesn't include Iran or North Korea. But it wasn't really about Iran or North Korea. It was about making America a less warlike, less intimidating, less—in a word—*imperial* nation.

In disdaining Britain, the president also disdains the things that Britain bequeathed to the Thirteen Colonies and, through them, the republic: the common law, a peculiar emphasis on personal freedom and property rights, distrust of government, a determination that laws should not be made, nor taxes levied, save by elected representatives.

To the anti-Anglosphere mind, there is nothing laudable about this patrimony. The main attraction of the American revolutionaries to Obama was not that they preached a set of sublime political ideals, but that they could be portrayed, after a fashion, as anticolonialists. He stressed this theme in his first inaugural address: "In the year of America's birth, in the coldest of months, a small band of patriots huddled by dying campfires on the shores of an icy river. The capital was abandoned. The enemy was advancing. The snow was stained with blood. . . ." He repeated it in a number of visits to developing nations. When the Arab revolts erupted in 2011, he compared the demonstrators to the Boston Tea Partyers—a compliment he never paid to the actual antitax movements in the United States.

The idea that the early Patriot leaders were upholding an inheritance—that they were conservatives, not radicals—doesn't fit this portrayal. Yet I hope that I have by now convinced you that the Patriots were consciously defending a political philosophy that, at that stage in human affairs, had been expressed only in the English language; that, indeed, is why they called themselves Patriots. Accepting this truth, however, means accepting the special place of the men who made the republic. It means accepting that America is not a multiculti fusion that owes as much to the Arapaho as to Adams. It means accepting that the United States is part of a wider family of English-speaking nations.

For more than a century, the United States was a dependable member of the Anglosphere alliance. Although not always formally allied with Britain and its settler colonies, the United States knew instinctively which side it was on. The English-speaking peoples fought side by side against the totalitarianisms of the modern age: fascism, communism, Islamic fundamentalism.

Even when the Anglosphere powers didn't formally take up arms in a common cause, no one doubted their common sympathy. The British government tacitly backed the United States in its war against Spain in 1898, and the U.S. government quietly returned the favor two years later during the South African war. The core Anglosphere states fought together in the two world wars, in Korea, in Iraq, and in Afghanistan. Only once, during the 1956 Suez Crisis, did the relationship break down—with, as it was to turn out, calamitous results. Eisenhower later called his failure to back Britain's military action against Gamal Abdel Nasser his "greatest regret."

When Argentina invaded the Falkland Islands in 1982, President Ronald Reagan knew who his friends were. He instructed the Defense Department to make available any matériel the British needed. Indeed, he went about as far as a benevolent neutral can go without formally opening hostilities, offering intelligence and logistical support to the United Kingdom. As Charles Moore showed in his biography of Margaret Thatcher, Reagan saw the ideological kinship between the Anglosphere states as more important than the fight against communism in South America. (Happily, in the event, the defeat of the Galtieri junta led to democracy in Argentina, not Marxism.)

Where Reagan saw an English-speaking democracy threatened by a dictatorship, Obama seems to see an imperial relic. Yet if the Falkland Islands are an imperial relic, a territory that just happened to be settled by English-speakers with a stubborn attachment to personal freedom and parliamentary rule, what is the United States? The Founders grew up with a strong sense that they were the heirs and guardians of a series of defined British freedoms. When they saw those freedoms arbitrarily withheld by a remote monarchy, they took to the battlefield to assert their rights. The liberties that the Patriot leaders believed they had won by inheritance were then encoded in a new constitutional dispensation. That is the essence of the American exceptionalism—the exceptionalism whose very existence the current president seems not to accept.

When Barack Obama sent back the bust of Winston Churchill that had stood in the White House, he was repudiating the foremost advocate of the idea that the Anglosphere nations have a special dream and a special task: that it is our duty, when others falter, to defend individual liberty, parliamentary supremacy, and the rule of law.

The current U.S. administration is not just anti-British. It has dropped the alliance that George Bush made with India, and even managed to pick fights with Canada over drilling rights. At the same time, President Obama has gone out of his way to cozy up to the EU. He more or less began his first presidential campaign with a European tour, whose climax was a rally in Berlin, where he delighted EU leaders by telling them that America had been too arrogant, that it could learn from them, and that he would do everything in his power to promote deeper European integration.

"It cannot be to our interest that all Europe should be reduced to a single monarchy," wrote Thomas Jefferson. But Barack Obama apparently knows better. He doesn't bother to explain how a European superstate is in American interests. Indeed, since the end of the Cold War, the EU has scarcely troubled to hide its anti-Americanism. It is building a satellite system with China, designed explicitly to challenge what Jacques Chirac called the "technological imperialism" of America's Global Positioning System. It refuses to deal with the anti-Castro opposition in Cuba, and declares U.S. sanctions to be illegal. It funnels cash to Hamas—now having to do so through nongovernmental organizations (NGOs) so as to get around its own rules on terrorist funding. It criticizes Washington on a whole series of essentially domestic issues: fuel prices, capital punishment, Guantánamo.

So why does the present administration suck up to Brussels? What's the attraction? I'm afraid it's hard to avoid the conclusion that Barack Obama wants the United States to be a lot more like the EU: highly taxed, eco-conscious, semipacifist, redistributive, centralized, indebted.

The key word in that list is *centralized*. When powers shift from people to government, from the states to the center, from elected representatives to bureaucrats, all the other things follow. The state machine becomes bigger, more extravagant, and yet, oddly, less effective. Jefferson's prescient analysis applies with equal force to Europe and America:

> Our country is too large to have all its affairs directed by a single government. Public servants, at such a distance, and from under the eye of their constituents, must, from the circumstance of distance, be unable to administer and overlook all the details necessary for the good government of the citizens; and the same circumstance, by rendering detection impossible to their constituents, will invite the public agents to corruption, plunder, and waste.

Fortunately for Americans, the U.S. Constitution was drawn up according to Jeffersonian principles. Unfortunately for Europeans, they had to make do with the former French president, Valéry Giscard d'Estaing, who preposterously told the drafting convention in Brussels "This is our Philadelphia moment." Where Jefferson immortally promised his countrymen "life, liberty and the pursuit of happiness," the EU's Charter of Fundamental Freedoms guarantees Europeans' right to "strike action," "free healthcare," and "affordable housing."

In the short term, of course, these things are pleasant enough: state-funded universities, shorter working days, two-hour lunch breaks, long vacations, early retirement: no wonder Europeans boast that their quality of life is superior to that of Yankee drudges. The trouble is that, when people enter the workforce older, leave it younger, and in all probability spend the few years in between working for the government, no one is generating much wealth. There comes a point when the money runs out. Europe has reached that point.

Money, to be precise, has run out for ordinary Europeans, trapped in their stagnant economies. The effect of decades of legislation on workers' rights is that the EU has plenty of rights, but fewer and fewer workers.

Money has not run out, however, for the Eurocrats, nor for the swollen class of consultants, contractors, and rent-seekers sustained by Brussels spending. As national governments struggle to rein in spending at home, every penny they save is swallowed up by the EU, whose budget continues to grow at 3 percent a year.

In Brussels, as in Washington, politicians talk about such money being used to "stimulate the economy." Which, in a sense, it does. To cite just one recent example, the president of the European Commission, José Manuel Barroso, and the president of the European Council, Herman Van Rompuy, flew to the same summit in Russia on separate private jets, leaving Brussels within four hours of each other. Why have one private jet when you can stimulate the economy twice as hard? Why have one EU head of government when you can have two? Why make do with one set of functionaries when two parallel sets can generate twice as many regulations?

The United States hasn't yet reached the point that Europe has. Because of the foresight and patriotism of its Founders, it is starting from a better place. But the direction of travel is unmistakable: permanent deficits, federal czars, Obamacare, support for supranational institutions. Day by day, the United States is becoming more European—by which I mean more regulated, more apologetic, more sclerotic, and more feeble.

Americans take pride in being self-reliant, optimistic, ambitious. But these characteristics are not a by-product of Mississippi water or turkey meat, and neither are they some magical quality in the American genome. People respond to incentives; culture is shaped by institutions. If taxation, spending, and borrowing keep rising, if more and more Americans become dependent on the state, it won't take long before they start behaving like the French, rioting and demonstrating in defense of their acquired entitlements.

Margaret Thatcher's political godfather, Sir Keith Joseph, used to remark that if you give people responsibility, they behave responsibly. What goes for individuals goes for entire nations.

For three decades, Greece was the single largest per capita net recipient of EU spending. Every year, the best and brightest graduates made the rational decision to work, either for the EU, for one of its associated bodies, or for the Greek state. The income, the gold-plated pensions, the tax perks were beyond anything they might realistically enjoy in the private sector. No one wanted to make anything or sell anything.

The results are visible in Greece today. Just as an individual can be infantilized by external subsidy so can an entire electorate. Asked whether they are prepared to suffer deeper cuts in order to remain in the euro, Greek voters reply, as a child might reply, "We want the euro, but we don't want the cuts."

When he was asked whether he believed in American exceptionalism, Obama replied: "I believe in American exceptionalism, just as I suspect the Greeks believe in Greek exceptionalism." At the time, I thought it an odd choice of analogy. Not anymore.

What, then, do we mean by American exceptionalism? What are the peculiar features of Anglosphere civilization that find their purest and freest expression in the United States? When we list them baldly, they can sound almost platitudinous. Free speech, free contract, free assembly, ownership rights, parliamentary control of the executive.

Yet these values, which have come to seem banal, are anything but assured. Free speech is now formally restricted in every Anglosphere nation except the United States—and, even there, it is under pressure. Since the early 1990s, numerous laws have criminalized various forms of opinion on grounds that they might cause offense to someone—typically someone from a racial or religious minority.

More recently, these laws have led to prosecutions. In 2007, Mark Steyn was taken to court in Canada for, in effect, writing disparaging things about Muslims. In 2011, an Australian commentator, Andrew Bolt, was convicted for writing an article in which he took a swipe at what he called "professional" Aborigines, by which he meant people of largely nonindigenous descent who identified as Aboriginal Australians for reasons of personal advantage. Although no one claimed that the article constituted a form of incitement—which has always been a crime under common law—it was nonetheless deemed to be illegal because of the offense it might cause, and because it had contained errors. In 2013, the United Kingdom, which had had an uncensored press since 1695, formally brought in a system of state regulation.

Such laws are nothing special in most of the world. Most European countries had always declared certain opinions—Holocaust denial, for example—to be criminal. In 2009, the United Nations Human Rights Council—on which Saudi Arabia sits alongside China and Russia—solemnly demanded that member states pass laws against the "defamation of religion." (It did not demand an end to persecution on grounds of religion, of course: that would have been awkward for China and Saudi Arabia.) The curious thing is that censorship came to the Anglosphere at the very moment when it had won. Throughout the Cold War, there was absolute freedom of speech in most English-speaking countries. Not until they were freed from the menace of communism did the Anglosphere states abandon one of the principles for which they had been fighting.

Before 1989, people in English-speaking states liked to tell each other that, unlike the poor wretches behind the Iron Curtain, they couldn't have their collar grabbed by a police officer for saying the wrong thing. Yet they are now regularly arrested for such offenses as quoting Bible verses that might offend gay people, or being rude about jihadi extremists. In one especially preposterous case during the 2009 British general election, a man who was sick of being canvassed by political candidates put a sign in his window reading, "Get them all out!" only to be visited by the police, who chose to interpret the words as an offensive remark about immigrants.

I wish I could tell you that such events were unique, but the truth is that they are becoming commonplace. A pianist in my constituency was investigated by the police because, at a dinner dance, he had played "Kung-Fu Fighting" in the hearing of a Chinese couple.

These sudden restrictions on free speech have come about as a result of another development that followed from the end of the Cold War, namely the elevation of human rights charters, national and especially international, over democratic legislatures.

Again, it is important to recall the difference between the Anglosphere conception of civil liberties—that is, as inherited rights won by our ancestors at an identifiable moment and passed down in the form of an inheritance—and the European notion that rights are bestowed by governments. While the *contents* of the various European and international human rights codes are generally unobjectionable, the elevation of such codes to the detriment of national sovereignty and parliamentary democracy is wholly at odds with the Anglosphere tradition, neatly summarized by Jefferson: "Our liberty can never be safe but in the hands of the people themselves."

As John Fonte showed in his magisterial study, *Sovereignty or Submission*, the rise of supranationalism has been sudden, and largely undebated. In the space of two decades, international law has moved from concerning itself with essentially cross-border issues, such as the status of diplomats and maritime rights, to essentially behind-border issues, such as labor law and the rights of minorities.

Fonte looks in detail at the way in which NGOs, unable to get their agenda through national parliaments, turn instead to international conventions. They are, indeed, quite overt about this. The United Nations conference on racism at Durban, for example, was held at the initiative of a number of leftist pressure groups that called openly on the UN to implement policies that had been rejected at the ballot box.

The trouble with arguing against international jurisdiction is that you can look as though you are on the side of some pretty unsavory characters: Omar al-Bashir, Radovan Karadzic, Slobodan Milosevic. If these brutes won't get justice in their own countries, runs the argument, surely they should get justice *somewhere*. Yet we are undermining, almost without thought, the traditional basis of the international order.

Territorial jurisdiction has been a remarkably successful concept. Ever since the Treaty of Westphalia in 1648, it has been broadly understood that crimes are the responsibility of the state where they are committed. Untune that string and hark what discord follows! Western liberals might say: "Since Karadzic won't get justice in Serbia, he should get it at The Hague." But an Iranian judge might apply precisely the same logic and say: "Adulterers in Western countries are going unpunished: we must kidnap them and bring them to a place where they will face consequences."

International jurisdiction breaks the link between legislators and law. Instead of legislation being passed by representatives who are, in some way, accountable to their populations, laws are generated by international jurists. We are, in other words, reverting to the premodern notion that lawgivers should be accountable to their own consciences rather than to those who must live under their rulings.

In consequence, as the late Robert Bork argued in *Coercing Virtue: The Worldwide Rule of Judges*, an agenda is being advanced that has been rejected at the ballot box. Courts make tendentious and expansive interpretations of human rights codes that go well beyond what any reasonable person would take the text to mean.

To pluck a random example, an illegal immigrant to the United Kingdom overturned his deportation order on grounds that he would not be able to access the same health care in his country of origin. The challenge was brought under Article 3 of the European Convention of Human Rights, which prohibits torture. It was a blatant example of an increasingly common tendency, namely the readiness of judges to rule on the basis of what they think the law ought to say rather than what it says.

With no meaningful scrutiny, international lawyers are able to suit themselves, meandering their way through gargantuan budgets, changing their own rules when they become inconvenient. As John Laughland showed in his study of the Milosevic trial, the International Criminal Tribunal on Yugoslavia admitted hearsay evidence, repeatedly amended its rules of procedure, and, when the old brute proved surprisingly eloquent in his own defense, took the extraordinary step of imposing counsel on him. Eight years and $200 million later, with the court no closer to a verdict, both judge and defendant were dead.

Then there is the sheer presumption of the new order. Indicting a head of state—as the International Criminal Court did in 2009 when it served a writ against the Sudanese president—amounts to declaring a war that one has no intention of fighting. The only way to bring President Bashir to trial would be to conquer his country and transfer sovereignty from him to the occupying powers: the basis of the Allies' jurisdiction at the Nuremberg trials. Without such a determination, international arraignments are declamatory: a way for those who serve them to feel good about themselves, even though their practical effect is to make tyrants dig in more deeply.

This brings us back to the main objection. While tyrants ignore international rulings, democracies—or, more precisely, judges *within* democracies—don't. Courts in Western countries increasingly use international conventions to challenge the decisions of their elected governments. For example, four successive British home secretaries tried to repatriate a group of Afghan hijackers who had diverted a scheduled flight at gunpoint to England's Stansted Airport. Despite the nature of their crime, and despite the removal of the Taliban regime from which they claimed to be fleeing, they were granted leave to remain in the United Kingdom through, in effect, judicial activism.

The politicization of international jurisprudence seems always to come from the same direction: a writ was served against Ariel Sharon, but not against Yasser Arafat. Augusto Pinochet was arrested, but Fidel Castro could attend international summits. Donald Rumsfeld was indicted in Europe, but not Saddam Hussein.

In 2010, Tzipi Livni, the centrist Israeli politician, was unable to attend a meeting in the United Kingdom because of a writ mischievously moved against her. The British government responded by changing its statutes so that the attorney general might strike down writs of a politically sensitive nature. Had this happened in any other context, moderate opinion would have been outraged. Imagine that, say, Robert Mugabe decided that one of his ministers would arbitrarily decide which foreign leaders might be hauled before the Zimbabwean courts. (Mugabe, come to think of it, is another leader whom the international human rights crowd seems not to have got round to indicting.)

Such absurdities follow inevitably from the abandonment of the tried and understood concept of state sovereignty, which operated effectively enough between 1648 and the 1990s. When was the internationalization of jurisdiction agreed? When was it even discussed? To quote Judge Bork again: "What we have wrought is a coup d'état: slow-moving and genteel, but a coup d'état nonetheless."

Legislatures have lost ground not only to activist courts but also to the standing bureaucracy of the state. Executive agencies and bodies have proliferated throughout the Anglosphere. In Britain they are known as "quangos": Quasi-Autonomous Non-Governmental Organizations. In the United States, their leaders have a name that at least accurately conveys their autocratic nature: federal czars.

In every English-speaking democracy the executive, in the form of various agencies with their alphabet soup of acronyms, has grown to a degree that earlier generations would have found incredible. If the Levelers could somehow be transported to our own time, and see the extent to which power in Britain is concentrated in the hands of the Child Support Agency, the Health and Safety Executive, the Food Standards Agency, and the rest, they would conclude that their achievements had been overturned. They had campaigned for popular control over what, in their day, were called Crown placemen. Now the placemen are back as agents, not of the Crown (although, in Britain and the Dominions, that remains the notional source of their authority) but of the cradle-to-grave interventionist state.

The state has grown as power has been centralized. We have seen that the dispersal of power was traditionally a distinguishing feature of Anglosphere political organization. From the seventeenth century onward, while European governments were ironing out local particularisms and centralizing power in their courts, the Anglosphere combined a uniform application of the law with a devolution of decision making unmatched anywhere else except Switzerland. While other states were concentrating power in the hands of royal commissioners, the Anglosphere retained its system of autonomous and accountable local government, complete with selectmen, justices of the peace, sheriffs, and the rest. Many Americans believe that these institutions were developed by the early Yankees; in fact, as David Hackett Fischer showed in *Albion's Seed*—one of the outstanding works of history of the late twentieth century—those offices, and the culture that sustained them, had been carried to New England from Old England, particularly East Anglia.

The English system of localism was, as Tocqueville observed, amplified in North America. Jefferson's fear that too large a distance between government and governed would "invite the public agents to corruption, plunder, and waste" actuated almost all early American statesmen. As so often, they were right.

One of the fixed laws of politics is that small units of government are more efficient than large ones. As decisions are taken further from the people they affect, waste, duplication, and producer-capture become more common.

Federation was the most common form of democratic organization in the Anglosphere. Australia, Canada, South Africa, parts of the Caribbean, India, Malaysia, and the United States all follow the federal model. Excluding very small Anglophone states, only New Zealand and the United Kingdom are unitary.

Yet, in all the federal nations, power has shifted steadily from the peripheral authorities to the center. There are reasons why this happens. Federal constitutions are interpreted by a supreme court, whose members tend to be appointed centrally, and so tend to have a national outlook. Indeed, the only Anglosphere state where centralization was retarded, and which consequently remained more federal than it would otherwise have done, was Canada prior to 1982: the British privy council, which was until then the supreme authority, had no particular interest in strengthening Ottawa vis-à-vis the provinces.

At the same time, state or provincial executives will often refer a decision upward in order to get around the opposition of their local assembly—only to find that, once transferred, the power cannot easily be regained.

The same is true of centralization that happens in an emergency. Powers that are passed to the central authorities on a supposedly contingent basis are hard to recover when the crisis passes.

War was the greatest driver of big government in the Anglosphere. The relative loss of freedom in the United Kingdom, relative to other Anglosphere states, dates from the Second World War, which consumed Britain more totally than its kindred nations. It wasn't the 1945 Labour government that created the welfare state, that Saturn that now devours its children. The real power grab came in 1940. With Britain's manpower and economy commandeered for the war effort, it seemed only natural that ministers should extend their control over health care, education, and social security. Hayek chronicled the process at first hand: his *Road to Serfdom* was published when Winston Churchill was still in Downing Street.

Churchill had become prime minister because he was the Conservative politician most acceptable to Labour. In essence, the wartime coalition involved a grand bargain. Churchill was allowed to prosecute the war with all the nation's resources while Labour was given a free hand to run domestic policy.

The social-democratic dispensation that was to last, ruinously, for the next four decades—and chunks of which are rusting away even today—was created in an era of ration books, conscription, expropriations, and unprecedented spending. The state education system, the National Health Service, the welfare state—all were conceived at a time when it was thought unpatriotic to question an official, and when almost any complaint against the state bureaucracy could be answered with "Don't you know there's a war on?"

Though they were not as badly affected as the mother country, Britain's overseas allies also experienced a surge in the power of the state during the two world wars.

In Canada, for example, direct taxation was a provincial prerogative until the First World War. The federal government imposed a direct corporation tax in 1916 and an income tax in 1917, and began to collect it directly, from most provinces, in 1941. In Australia, the first federal income tax was introduced in 1915, and the central government consolidated its control of all income tax in 1942. As a proportion of Australia's GDP, taxation doubled during the First World

War (from 5 to 10 percent), and doubled again in the Second (from 11 to 22 percent).

In the United States, too, the engorgement of government was a consequence of centralization. Here, though, the main driver was not war but the massive power grab instituted by the Roosevelt administrations from 1933. Toward the end of his distinguished life, the great economist Milton Friedman looked back on the changes through which he had lived:

> From the founding of the Republic to 1929, spending by government at all levels never exceeded 12 percent of national income except in times of major wars, and federal spending typically amounted to 3 percent or less. Since 1933, government spending has never been less than 20 percent of national income and is now over 40 percent.

And that was before the 2008 financial crash and the splurge that followed.

These tax levels would have prompted revolution among earlier Anglosphere generations. In 1900, a British household typically spent 8.5 percent of its income on government—a figure little changed since the days of the medieval tithe. In the United States, the equivalent figure was 6.5 percent.

Now the proportions have risen to 46 percent in the United Kingdom and 36 percent in the United States. Tax is by far the largest item of household expenditure. It accounts for far more than almost any working family spends on its mortgage, its car, or its bills.

And even those statistics don't tell the whole story. Tax levels have reached saturation point, but spending continues to rise. In order to cover the differences, Western governments are borrowing from what Shakespeare called "your children yet unborn and unbegot." The average American earns $70,000 a year, making the

United States, to outward appearances, a very wealthy country. Set against that salary, though, is each American's share of public debt: $135,000, on top of whatever private debts he has incurred. Paying the interest on government debt will cost the average U.S. citizen $11,000 a year.

Those figures, in a sense, are all we need to know about the present crisis. The Anglosphere nurtured a unique political culture in which the individual was larger than the state, and in which there were mechanisms to hold the government to account.

Those mechanisms are no longer working. The state machine has outgrown democratic scrutiny. Agencies and executive bodies proliferate beyond the purview of elected representatives. Like Asimov's robots, they have learned to program each other without human intervention.

Budgets are, in practice, set by members of the executive, who have an immediate personal interest in higher spending, rather than by representatives of the legislature, who represent taxpayers. The Anglosphere's political model is beginning to look a lot like anyone else's: like that of the Mings or the Moguls or the Ottomans or the Incas or the Carolingians or any other group who, once in a position to expropriate their citizens with force of law, do so at will: what the nineteenth-century French philosopher Frédéric Bastiat called "legal plunder."

It was the breaking of that pattern in the Anglosphere, especially after 1689, that led to the extraordinary prosperity that followed. Monopolies and guilds were replaced by competitive businesses. Modern notions of free trade and open competition were evolved. Here, for example, is Nicholas Barbon, writing in 1690:

> The Prohibition of Trade, is the Cause of its Decay; for all Forreign Wares are brought in by the Exchange of the Native: So that the Prohibiting of any Foreign Commodity, doth hinder the Making

and Exportation of so much of the Native, as used to be Made and Exchanged for it. The Artificiers and Merchants, that Dealt in such Goods, lose their Trades; and the Profit that was gained by such Trades, and laid out amongst other Traders, is Lost. The Native Stock for want of such Exportation, Falls in Value, and the Rent of the Land must Fall with the Value of the Stock.

Lower tariffs led to specialization, comparative advantage, and unprecedented rise in wealth. Security of contract led to confidence in credit. Public opinion set a limit on levels of taxation. Individual liberty encouraged a culture of enterprise. Social mobility rewarded effort. After hundreds of thousands of years of economic flatlining, our species took off. We need only look around us to see how far we are drifting from the principles that made the take-off happen.

Conclusion: Anglosphere Twilight?

We have traveled a long way since the gloomy forests of first-century Germany. We have seen that the story of the English-speaking peoples is the story of how they imposed their will upon their rulers. We have noted the way the primitive tribal meetings of the early Teutonic peoples evolved into the local assemblies of the settlers in England, into the Witans of the Anglo-Saxons, and, after many fierce struggles, into the direct ancestor of the parliamentary bodies that meet throughout the Anglosphere today.

We have observed, too, the brave role played by the common law: that beautiful, anomalous system that belongs to the people, not the state, and which allows criminal justice and civil disputes to be domesticated. We saw the common law create the features that kept the English-speaking peoples free, from jury trials to habeas corpus. We looked on as it, and the values it inspired, overthrew the Stuart tyranny and made the American republic. We watched it serve as an antibody against the infections of slavery and dictatorship.

We have seen an idea that had its roots in theology—the notion that every individual must answer for himself, without the intervention of priests or prelates—pass into political theory. We have watched the spores of that idea break away from their religious roots. We have gazed as the wind carried them to far lands. We have seen them take root throughout the Anglosphere in the form of the doctrine of personal responsibility.

And we have seen how a happy accident of nature made possible the triumph of constitutional liberty. Because Great Britain was an island, and the Anglosphere a form of extended archipelago, there was no need for a permanent standing army. Taxes were commensurately low, and the government commensurately weak. If the regime needed resources, it had to collect them by consent through the people's representatives.

We have seen, finally, how the peculiarities of English property law, based on individual rather than familial rights and on primogeniture, sustained the individualist culture that was in time to develop into capitalism as we now know it: that is, a system where every individual is free under the law to sell his services through private contracts. We have grasped the essence of the Anglosphere miracle: the move, as Sir Henry Maine put it, from status to contract that is the ultimate guarantor of a free economy and a free society.

How, having taken in this much, could we not be disquieted by the readiness with which our generation has squandered its heritage?

That abandonment has gone furthest in the United Kingdom. As it has surrendered its sovereignty to the EU, so it has progressively surrendered the various elements of its national distinctiveness. Laws are passed by European Commissioners, who are appointed, not elected. Trade has been artificially redirected from the English-speaking hinterland to the cramped and dwindling European customs union. Power has shifted from Parliament to the standing apparat, both at national level and in Brussels. The connection between taxation and representation has been broken, as the EU has acquired direct revenue-raising powers. The state has swollen to a previously unimaginable size. Even the common law itself, the first and last bulwark of Anglosphere liberty, is being battered down.

In the 1970s, Lord Denning, the former Master of the Rolls and greatest modern English jurist, likened EU law to an incoming tide, swelling England's river mouths and estuaries. In 1990, toward the end of his extraordinary life, he revised the metaphor: "Our courts must no longer enforce our national laws. They must enforce Community law. No longer is European law an incoming tide flowing up the estuaries of England. It is now like a tidal wave bringing down our sea walls and flowing inland over our fields and houses—to the dismay of all."

In the United Kingdom, the process of Continentalization is direct and tangible, driven by force of law. In the rest of the Anglosphere, it is indirect, and its motors are cultural rather than legal. There has been a general loss of confidence in the superiority of the Anglosphere model, which fended off every extremist challenge throughout the twentieth century. Cultural relativism feeds into hard policy. Once you reject the notion of exceptionalism as intrinsically chauvinistic, you quickly reject the institutions on which that exceptionalism rested: absolute property rights, free speech, devolved government, personal autonomy. Bit by bit, your country starts to look like everyone else's. Its taxes rise; its legislature loses ground to the executive and to an activist judiciary; it accepts foreign law codes and charters as supreme; it drops the notion of free contract; it prescribes whom you may employ and on what terms; it expands its bureaucracy; it forgets its history.

Is it any wonder, as this process unfolds, that power is shifting? The maps of the world that we know, centered on the Atlantic Ocean, no longer reflect the geopolitical reality. The economic center of gravity is moving fast. In 1950, it hovered over the Atlantic Ocean, off Iceland. In 1980, it had moved closer to Norway. Today, it is speeding east across the frozen tundra of Russia. Ten years from now, it will be just beyond the northeastern frontier of Kazakhstan.

This is partly because the Asian states have liberalized. China and Russia, though they remain closed autocracies, are no longer as totalitarian as they were in the 1980s. India has opened its economy impressively.

At the same time, though, the English-speaking states are going in the opposite direction, pursuing the Ming-Mogul-Ottoman road to uniformity, centralization, high taxation, and state control. No wonder they are losing their preeminence.

There is nothing inevitable about this process. It is our choice, not our fate. The fault, dear Brutus, is not in our stars, but in ourselves, that we are underlings.

What, then, is the remedy? After so long a disquisition on the nature of Anglosphere exceptionalism and its enemies, the answer may sound so curt as to be glib, but it is no less heartfelt for that. We should remember who we are.

The basis of the Anglosphere is its common values and institutions, not the formal links between its governments. To the extent that such links matter at all, we should not aim for anything supranational or, indeed, anything that expands the power of the state. We should aim, rather, for an Anglosphere free trade area.

The United States and Canada currently form a single market as, in most regards, do Australia and New Zealand. The United States and Australia signed a free trade agreement in 2005, and New Zealand is negotiating one as I write. Singapore's eighteen free trade agreements include treaties with the United States, Australia, and New Zealand. India is starting from further behind—a legacy of the postcolonial period—but is liberalizing quickly. The main problem lies with the United Kingdom and Ireland, which, as members of the EU, cannot sign independent commercial agreements, but are instead held back by Brussels protectionism.

If these two states were outside the EU, an Anglosphere free trade area could be based on the unhindered movement of goods, services, and capital; and on an easing, if not a complete lifting, of restrictions on the free movement of labor.

It is fitfully suggested that the Commonwealth, which is growing at an impressive speed, could be a suitable vehicle for such economic integration. The trouble with the Commonwealth, though, is that it contains some dictatorial regimes that have drifted away completely from Anglosphere values; and, conversely, it doesn't include two key component states: the United States and Ireland.

Better to begin with the key relationship of our present age, namely that between the core Anglosphere states and India. The days when India pursued an essentially anti-Western foreign policy through its leading role in the Non-Aligned Movement are over for good. The country is now a major military ally of both the United Kingdom and the United States, though the present U.S. administration has not pursued the relationship with the same warmth as its predecessor.

It was the military aspect of the Anglosphere that actuated its greatest champion of all. In 1946, in the tiny town of Fulton, Missouri, Winston Churchill gave a speech remembered to this day for a sentence where he spoke of an iron curtain having descended across Europe, from Stettin in the Baltic to Trieste in the Adriatic. Yet that line was almost incidental. Churchill was quite clear about what his chief purpose was:

> I come now to the crux of what I have travelled here to say. Neither the sure prevention of war, nor the continuous rise of world organization will be gained without what I have called the fraternal association of the English-speaking peoples. This means a special relationship between the British Commonwealth and Empire and the United States of America.

This was no mere rhetorical flourish. Churchill knew that the survival of liberty had been secured by an Anglosphere military victory. He had played a brave role in all three of the great twentieth-century conflicts that pitted the English-speaking peoples against their autocratic foes, fighting in the first, inspiring the tribe in the second, defining the third. And he knew exactly what he wanted, namely a permanent, formalized military alliance. One, indeed, that would go beyond any alliance then known between sovereign states:

> Ladies and gentlemen, this is no time for generality, and I will venture to the precise. Fraternal association requires not only the growing friendship and mutual understanding between our two vast but kindred systems of society, but the continuance of the intimate relations between our military advisers, leading to common study of potential dangers, the similarity of weapons and manuals of instructions, and to the interchange of officers and cadets at technical colleges. It should carry with it the continuance of the present facilities for mutual security by the joint use of all Naval and Air Force bases in the possession of either country all over the world. This would perhaps double the mobility of the American Navy and Air Force. It would greatly expand that of the British Empire forces and it might well lead, if and as the world calms down, to important financial savings. Already we use together a large number of islands; more may well be entrusted to our joint care in the near future.

It was not simply an Anglo-American alliance that Churchill wanted. He understood that the Anglosphere went further than those two core states:

> The United States has already a Permanent Defense Agreement with the Dominion of Canada, which is so devotedly attached to

the British Commonwealth and the Empire. This Agreement is more effective than many of those which have been made under formal alliances. This principle should be extended to all the British Commonwealths with full reciprocity. Thus, whatever happens, and thus only, shall we be secure ourselves and able to work together for the high and simple causes that are dear to us and bode no ill to any.

He ended with an aspiration that has been almost forgotten today: "Eventually there may come—I feel eventually there will come—the principle of common citizenship, but that we may be content to leave to destiny, whose outstretched arm many of us can already clearly see."

What would the great man think if he could be transported to our present age? Much of the common defense infrastructure he called for, and indeed helped put into place during his two terms in office, remains. The five core Anglosphere states retain a closeness on issues of military technology, including nuclear technology, that has no equivalent among separate states anywhere on earth. They operate a joint electronic eavesdropping system known as Echelon, sharing secrets under a treaty known as UKUSA, which dates back to 1947. This much, we may be sure, would merit a grunt of unsurprised approval from that deep chest.

What would astonish the old war leader, however, would be to find that India was now the second-largest investor in the United Kingdom after the United States, that its economy was poised to overtake Britain's, that it was a nuclear power, and that it had achieved all these things while remaining a law-based democracy.

Once he had overcome his astonishment, he would surely press for the immediate inclusion of India into the military and intelligence structures of the Anglosphere. He would recognize India as a state that met the criteria he saw as essential to civilization, namely a mechanism to change the government peacefully, a legal system that is independent of the rulers, and a conception of property rights that protects the freedom of the individual.

Nor would India be his only surprise. He would doubtless shake his round and stubborn head in wonder at the way Singapore had been transformed from a mosquito swamp into a gleaming city-state. He would nod approvingly at the way the more democratic Caribbean states had retained their maces and their horsehair wigs and their stiff blue passports. And he would surely take particular delight in the way that South Africa, despite the very different future envisaged by his former jailers the Afrikaners, had rejoined the Commonwealth as a democracy.

One thing, we may be certain, would leave him bewildered and depressed. He would struggle to explain the loss of confidence in that portion of the English-speaking peoples whose history he had written, namely the core Anglosphere states. He would wonder why, having seen off the authoritarian challenges of both fascism and Marxism, they were now so ready to discard the things that had raised them to greatness. Still, he was fundamentally an optimist, and would doubtless deliver some well-turned quip about doing the right thing after exhausting all other possibilities.

For we are not finished. We remain an inventive, quizzical, enterprising people. All we need to do is hold fast to the model that made us that way. Edmund Burke's words about America in 1775 apply, mutatis mutandis, to the Anglosphere as a whole today. English privileges have made it all that it is; English privileges alone will make it all it can be.

At almost exactly the moment that Edmund Burke was making that speech, at the other end of the Anglosphere, a young doctor in Boston named Joseph Warren—the man who sent Paul Revere on his ride—was seeking to rally his countrymen in defense of the same principles. His words ring down the ages: "You are to decide the question on which rest the happiness and liberty of millions yet unborn. Act worthy of yourselves."

You, reading these words in his language, are the heirs to a sublime tradition. A tradition that gave us liberty, property, and democracy, and that raised our species to a pinnacle of wealth and happiness hitherto unimaginable. Act worthy of yourselves.

Index

About the Author

DANIEL HANNAN is a writer and blogger, and he has been a member of the European Parliament representing South East England for the Conservative Party since 1999. He graduated with a double first in history from Oriel College, Oxford, and worked as a speechwriter and journalist before standing for election.

He is a leading Euro-skeptic and is secretary-general of the Alliance of European Conservatives and Reformists, which brings together center-right parties from around Europe committed to national sovereignty and free markets.

His previous book, *The New Road to Serfdom: A Letter of Warning to America*, was a *New York Times* bestseller.

He blogs at www.hannan.co.uk.

About the Author

DANIEL HANNAN is a writer and blogger, and he has been a member of the European Parliament representing South East England for the Conservative Party since 1999. He graduated with a double first in history from Oriel College, Oxford, and worked as a speechwriter and journalist before standing for election.

He is a leading Euro-skeptic and is secretary-general of the Alliance of European Conservatives and Reformists, which brings together center-right parties from around Europe committed to national sovereignty and free markets.

His previous book, *The New Road to Serfdom: A Letter of Warning to America*, was a *New York Times* bestseller.

He blogs at www.hannan.co.uk.